The Internet, Power and Society

CHANDOS
INTERNET SERIES

Chandos' new series of books are aimed at all those individuals interested in the internet. They have been specially commissioned to provide the reader with an authoritative view of current thinking. If you would like a full listing of current and forthcoming titles, please visit our website www.chandospublishing.com or e-mail info@chandospublishing.com or telephone number +44 (0) 1223 891358.

New authors: we are always pleased to receive ideas for new titles; if you would like to write a book for Chandos, please contact Dr Glyn Jones on e-mail gjones@chandospublishing.com or telephone number +44 (0) 1993 848726.

Bulk orders: some organisations buy a number of copies of our books. If you are interested in doing this, we would be pleased to discuss a discount. Please e-mail info@chandospublishing.com or telephone number +44 (0) 1223 891358.

The Internet, Power and Society: Rethinking the power of the internet to change lives

MARCUS LEANING

Chandos Publishing

Oxford · Cambridge · New Delhi

Chandos Publishing
TBAC Business Centre
Avenue 4
Station Lane
Witney
Oxford OX28 4BN
UK
Tel: +44 (0) 1993 848726
E-mail: info@chandospublishing.com
www.chandospublishing.com

Chandos Publishing is an imprint of Woodhead Publishing Limited

Woodhead Publishing Limited
Abington Hall
Granta Park
Great Abington
Cambridge CB21 6AH
UK
www.woodheadpublishing.com

First published in 2009

ISBN:
978 1 84334 452 0

© Marcus Leaning, 2009

British Library Cataloguing-in-Publication Data.
A catalogue record for this book is available from the British Library.

Typeset in the UK by Concerto.

Contents

List of tables

About the author

Marcus Leaning is originally from London. He is a senior lecturer and programme leader on the media studies programme in the School of Media and Film at the University of Winchester. Prior to moving to Winchester he taught at Trinity College, Carmarthen, University of Wales for several years. He has also taught in Japan and Thailand and has been an IT trainer and web designer.

He was awarded his PhD in 2004 from the University of Bedfordshire in media sociology applied to new media. His teaching and research interests centre upon digital media, media theory, media literacy and the intersection of media technology and society. Marcus has published numerous chapters, articles and conference papers on these and other areas, and is the editor of a two-volume book on media and information literacy – *Issues in Information and Media Literacy: Criticism, History and Policy* and *Issues in Information and Media Literacy: Education, Practice and Pedagogy* (Informing Science Institute Press, 2009).

Acknowledgements

In writing this book I received the help and assistance of many people. Quite often people did not even know they were helping me, and it was only afterwards that I realised that a passing conversation had a lasting effect upon me and has contributed in some way to this text. I am afraid, therefore, that I can't even begin to list all the people who have assisted me in this endeavour. However, the people listed here do deserve a mention, and if I have neglected anyone it is simply a small error and I apologise most strongly. So, in no particular order...

I would like to thank the staff of the School of Creative Arts and Humanities at Trinity College, Carmarthen, for their time and willingness to support me with a partial sabbatical. In particular I wish to thank Paul Wright, Neal Alexander, Jeni Williams, Michelle Ryan, Ceri Higgins, Brett Aggersberg and Sarah Morse.

I would also like to thank my students on the Digital Cultures course, who have been exposed to, and helped to shape, the ideas in this book.

I wish to thank the staff of Chandos Publishing for publication, Cherry Ekins for copyediting and preparing the text and Hireascribe for working on the text.

I thank Richard Chamberlain, Paul McDonald and Will Merrin for help in preparing the proposal; and Claire

O'Neill, Shane Doheny, Chris Wigginton and Udo Averweg for their encouraging words.

I owe a particular debt to a number of people who, perhaps unwittingly, have considerably aided my intellectual development (for what it is): Rafiq Ghanty, Mirko Petric, Inga Tomic Koludrovic, Adrian Page, Garry Whannell, Alexis Wheedon, Julia Knight, Richard Wise, Peter Dean, Gavin Stewart, George Cairns and Dave Green.

Finally, I owe a large debt to my wife Sadhbh, my daughter Tara and my father Richard.

List of acronyms

BBS   bulletin board system

HCI   human-computer interaction

ICT   information and communication technologies

ICTD   information and communication technologies for development

MMORPG massive multiplayer online role-playing game

MUD   multi-user domain or dungeon

OSI   Open Source Initiative

OSM   Open Source Movement

Introduction

This book is about how the internet has been understood. More specifically, it is concerned with how the internet has been thought of as a technology that can do great things and change lives. Indeed, there have been numerous predictions as to how daily life will be transformed and (probably) improved through the use of the internet: in 2000 Bill Gates spoke of the internet's ability to invigorate economic systems (Gates, 2000); former US vice president and 2007 Nobel Prize laureate Al Gore views the internet as a technology that could empower citizens and lead to a revival of democratic process (Gore, 1994); and a UCLA report by Jonathan Cole et al. (2001) even indicated how the internet could contribute to social and psychological well-being. More recently, specific applications and practices such as social networking and blogs (weblogs) have been thought to transform politics and culture radically (Barlow, 2007; Kline, 2005; Scott-Hall, 2006). Conversely, the internet is also seen as a threat: it is a new channel for the transmission of paedophilic material; hackers can attack governments across borders and even attack computers in our homes; our identities can be stolen; and people become addicted to online communication and playing computer games online.

Such optimistic and pessimistic visions are not peculiar to the internet as a technology or to the age in general;

numerous predictions have previously been made as to the enticing and terrifying possibilities of new technologies. Marvin's fascinating study, *When Old Technologies Were New*, describes beliefs surrounding the emergence of two technologies during the late nineteenth and early twentieth centuries: the electric light and the telephone (Marvin, 1990). Marvin notes how the telephone was understood in different ways and attributed with numerous social consequences. Similarly Standage's *The Victorian Internet* explored how the telegraph was also viewed as both a wonderful development and a terrible threat (Standage, 1999). Contemporary discussions concerning the internet are in many ways similar. We see the internet in many different ways – sometimes as saviour, sometimes as a threat, sometimes as a beneficial force in society, bringing democracy and new opportunities, and sometimes as a modern curse, allowing illicit content within the reach of our children.

In attempting to deal with such new technology, answers have been sought as to how the internet, a media form that seems so potent and so very different from previous media technologies, should be understood, studied, used and managed. Moreover, in trying to understand the internet, interest comes from a wide range of sources. In addition to a large amount of popular comment concerning the internet, the subject has been of considerable interest to humanities and social scientific branches of academia. A wide range of theories have been deployed and developed to make sense of the internet's usage and its impact upon society. Within this field three approaches may be discerned.

An early though still influential approach emerged from technological disciplines such as computer science and engineering. This approach is concerned primarily with technical issues: how the computer operates and how the user and computer interact. From this perspective, the user is primarily regarded in a biological or cognitive sense, and notions of the social world are not usually considered. This is still a very influential approach and informs much technical discussion of the internet and the various ways in which the internet is conceptualised in technological circles.

Partially in response to this coldly scientific approach, specific forms of knowledge and epistemological systems drawn from long histories of understanding the social world have been deployed. Thus forms of social psychology (Wallace, 2001), psychoanalysis (Holland, 1996) and community sociology (Wellman et al., 2002; Kollock and Smith, 2001) have all been used in attempting to describe online activity and the consequences of such activity for users. These approaches have tended to relate existing theories directly to online communication. Attention is focused upon the user and the 'social space' of communication. However, in many ways such approaches ignore the technology, which is regarded as a background that barely affects the social activity occurring.

A further response contests that new approaches must be developed that incorporate a notion of the internet in the theoretical approach rather than simply applying existing theories to the internet. S. Jones (1998: x) proposes that 'simply applying existing theories and methods to the study of internet-related phenomena is not a satisfactory way to build our knowledge of the internet as a social medium'. Instead, Jones contends that we should develop forms of understanding that accommodate the unique nature of the internet – the internet is 'a medium that intersects with everyday life in ways that are both strange and omnipresent' (ibid.: ix–x). The internet should be considered a potent phenomenon, able to transform the environment in which it is used and be factored into theories at the most fundamental of levels. The internet should be the central focus and not regarded as being on the periphery – a passive yet enabling technology. It should be acknowledged and recognised in all accounts, and the way in which it changes our very forms of communication and social relationships should be the focus of study.

However, the issue is complicated further; the manner or way in which a technology or medium such as the internet is conceptualised is far from transparent or fixed. There is no single understanding of the internet, or even of technology in general. Numerous theories and readings circulate as to what is meant by technology and how it impacts upon individuals and society. Indeed, historically technology has, as Escobar (1994) notes, escaped the same rigorous examination within the humanities and social sciences that has been applied to other spheres of human activity and material production. Therefore, moving to questions such as a specific methodology prior to establishing what it is we are studying (or what we think we are studying) is problematic.

The ideas presented in this book contribute to an emerging sociological and critical perspective that challenges simplistic readings of the internet. As Selwyn and Gorard (2002: 6) propose:

> There is clearly a pressing need to step beyond the limitations of previous analyses of ICT if we are to gain a deeper understanding... We need to be aware of the social, cultural, political, economic and technological aspects of ICT – the 'soft' as well as the 'hard' concerns.

This perspective challenges uncritical and discrete readings of technology and society – it explicitly disputes the idea that technology and society are separate. This perspective lies broadly within the 'democratic' approach to technology and technological policy-making advocated by authors such as Sclove (1995).

Such a perspective is further divided into a number of positions differentiated by the broadness of focus. A number of studies have deployed in-depth anthropological methods studying micro-usage of the internet. For example, Miller and Slater (2000) explored the use of the internet in Trinidad. Similarly, the radical constructivism of authors such as Hine (2000) draws upon the critical positions on the social construction of scientific meaning within science and technology developed by Latour and Woolgar (1986). Authors such as Slevin (2000) advocate that attention should be directed at a broader vista and offer a 'social systems theory' of the internet. Attention is focused at the societal level, an approach that accepts notions of collective understanding and action, and it is at this level that this book partly operates.

In the concluding chapters of this book I advocate a particular standpoint for the study of the internet. I propose that its impact upon society and the way in which it changes lives occur in a particular manner. I argue that the internet intersects with 'social systems', and reorientates such systems in a particular manner. In doing so the internet facilitates and accelerates our experience of certain features of late-modern social life. However, the propensity to do so is dependent or contingent upon other factors. The internet may cause change and, as will be noted, may bring about new ways of acting in a political sense. However, its ability to cause change is deeply linked to other aspects of social life – it is contingent.

Theoretical foundations and general approach

As mentioned above, there have been numerous attempts to understand the widespread use of the internet. Such efforts are far from homogeneous in origin and come from a wide range of academic and theoretical traditions. Indeed, the current state of study of the internet is in many ways similar to the academic discipline of media studies; the study of the internet now constitutes a field of study rather than an agreed set of research practices and epistemological truths. Even within sociological and humanities-oriented studies of the media, a considerable number of approaches and traditions have arisen.[1] Such competing areas afford a wealth of ways in which to approach a subject and similarly innumerable opportunities for conflation of terminology and direction. While acknowledging the equally valid explanatory potency of differing positions and perspectives, it would prove an impossible task to do justice to all positions. It is useful, therefore, to declare a general theoretical position early on. The approach adopted here is broadly within contemporary sociological areas of concern. It draws heavily upon developments in social theory by, and in reaction to, Giddens (1976, 1979, 1984, 1990, 1991, 1994) and Beck (1992, 1994, 2007; Beck and Beck-Gernsheim, 1995, 2002, 2004; Beck and Lau, 2005; Beck et al., 2003). The particular aspect of Giddens's and Beck's work used is one concerned with the process of modernisation and the maturing of certain societies into 'late-modern' forms. Strongly present throughout this book (particularly in the later chapters) is the description of modernity provided by Giddens and Beck and a general orientation sympathetic to Giddensian and Beckian sociology.

Interestingly enough, neither Giddens nor Beck has made a systematic attempt to examine the internet or even the media in general (though there have been a number of attempts to build some aspect of their work into studies of the media – Thompson, 1990, 1995 – and the internet – Slevin, 2000). Instead, Giddens's and Beck's work tends to focus upon the far broader remit of changes in a number of contemporary Western societies. Specifically, certain aspects of their work have been concerned with how the individual in 'late-modern' life has undergone changes in the way in which identity is formed. Giddens and Beck contend that the way in which identity is formed is different in those societies with a 'late-modern' social form than in those of preceding 'modern' social forms. A number of factors peculiar to those societies experiencing a more intensified form of modernity subtly change the way in which individuals interact with each other and the social world. Late-modern societies are characterised by an intensification of the processes of 'detraditionalisation' and 'individualisation'. While such processes are present in those societies experiencing modernity, they are felt to a much greater degree in late-modern societies. It is this aspect that is made use of, particularly in later chapters.

Structure

This book is divided into two parts. The first part (Chapters 2–5) is a description of how the internet is generally understood. This part examines discussions that surround the internet and explores how such discussions operate on a number of levels. Chapter 2 explores debates surrounding the relationship of technology to society. Utilising a model developed by the philosopher Feenberg (1999a, 1999b,

2003, 2004), three forms of understanding technology's interaction with social life are examined. These interpretations are all, to some degree, present in contemporary discourse surrounding the internet. The recognition that such conflicting traditions exist – technology is not an absolute truth, but a field of discourse and conflicting theories – is a position that sociological studies advocate. Sociological approaches decentre technology in accounts of the operation of the internet and replace it with a recognition that technology and society are inextricably linked.

In spite of this intention to decentre technology, an account or description of the internet is needed. Consequently, Chapter 3 describes how the internet is conventionally understood. As noted above, much has been made of the internet's potential to enhance communication and invigorate a more active form of citizenship. Furthermore the 'advent of Web 2.0' – the new developments, platforms, applications and specific communicative practices associated with 'social networking' – has heightened awareness of the interactive qualities ascribed to the internet. Therefore this chapter explores the qualities that seem to differentiate the internet from older forms of media. However, this examination does not take the form of a simple description of the operation of the technology, as to do so reverts to a kind of technological essentialism – that the internet operates in a specific fashion. Instead, this chapter examines the general and specific understandings of the internet that inform discussion. Attention is paid to the ways in which the internet is described in journalism and academic literature in the humanities and social sciences. It is argued that, within such literature, a distinct belief that the internet possesses qualities that distinguish it from older forms of media operates. Therefore the focus of the chapter is on how a number of perceived features of the internet are understood and articulated: interactivity, interpersonal communication, the user production of disseminable information and the individualisation of media content. It is not proposed that such qualities are 'essentially' present in internet technology, nor that such qualities actually distinguish the internet from previous forms of media, merely that these are certain perceptions surrounding the internet.

Chapter 4 examines the how the internet is thought to result in new forms of social association. The unique qualities of the internet are believed to enable new patterns of communication between individuals and between the individual and media producers. These new forms of association have been understood as imbued with a distinctly political potential. Of particular note is the understanding that the internet contributes to the re-establishment of the public sphere and enables new forms of politics. Consequently, the internet is thought to be able to be used 'anti-systemically'. This potential has been conceptualised in a variety of ways, ranging from a corrective force acting for the re-establishment of democratic forums and a media form that may challenge the pathologic incursions of state or commerce into public discourse to a truly radical medium affording revolutionary potential.

The internet is, of course, not the only form of media to have been envisaged as possessing a potency to affect political life. Forms of mass media such as underground magazines and 'pirate' radio have often been deployed in a radical fashion. Chapter 5 examines a number of theories of alternative media, and addresses how the internet may be understood in such a way. Various pluralist understandings of the role of the media are juxtaposed with more conflict-

and critically oriented conceptions of the role of the media and the internet. It is noted that the conception of the internet as a form of political media is deeply rooted in Western European and American models of the media and politics.

This proposition, the reading of the internet as an implicitly political media form, owes much to particular models of politics and the media and illustrates a point made by the media theorist Downing. Downing (1996) argues that the majority of accounts of the media originate in a disconcertingly small number of 'laboratories', namely the USA, the UK and a few other Western European countries. The number of studies carried out concerning non-Western societies and countries is limited. When examining the political role that the media and the internet play in non-Western societies the number is comparatively even smaller. Existing understandings of the internet and models of its political role have emerged from studies in a small number of countries. Downing (ibid.: xi) warns that to use such a small range of countries to develop theories of the media is 'both conceptually impoverishing and a peculiarly restricted version of Eurocentrism'.

The second part of the book (Chapters 6 and 7) extends these arguments and develops a more general approach to the study of the internet that relates its use to the 'form' of a society. Chapter 6 examines how the 'form' of a society and its relationship to the processes of social change occurring with modernity considerably affect the potency of media. This argument uses a description of modernity, drawn from the work of Giddens and Beck, and incorporates the idea that a number of late-modern societies are undergoing a process of individualisation. Individualisation refers to an increasingly emergent trend, ethos or project within such societies in which patterns of identity formation increasingly centre upon the individual and away from traditional social systems. Strikingly, it is in the societies that most epitomise individualisation that the internet is most studied and in which it is seen to offer potential to liberate. Accordingly, it is proposed that societies with different societal emphasis upon ideas such as individualism may use or encounter technology or media in different ways. The political power of the internet, believed to be an implicit aspect of the technology, is only realised in particular social environments. The internet requires the correct social conditions for its full political potential to be manifest or evoked.

In Chapter 7 these arguments are developed into a sociological standpoint for the study of the internet. This standpoint is founded on two assertions: much contemporary theorising concerning the internet arises from a particular reading of the internet, a reading that owes much to a singularly Western conception of the role of media and democracy; and the way in which a medium functions is deeply linked to the social form of a society, and particularly the experience of the society with the passage of modernity.

Accordingly, the internet is not understood as detached from, or determining, society and social systems. Instead, media technologies such as the internet are best understood when regarded as deeply tied to social systems. The potential of the internet to operate as an enabler of new forms of politics is contingent upon the various conditions. The internet is a contingent form of communication; it operates in a certain way when certain conditions occur. Therefore, methodologically the internet is best understood by examining the relationship between the technology (and its social construction), the society in question and the form of action studied.

The approach of this book both acknowledges the common discourses that surround the internet and explores how its use needs to be understood as being deeply interlinked with social life. It is sociological, as it involves a commitment to placing an understanding of society and the social construction of the internet at the centre of the field of study.

Finally, the Conclusion briefly sketches how this argument could possibly be used. I firstly discuss the issue of developing actual methods of investigation, and secondly examine how the approach fits well with recent developments in two areas of academic study: the use of information and communication technologies for development, and the challenge that new media present to traditional media studies.

Notes

1 Raymond Williams (1981) initially made the argument that three traditions dominate media and cultural studies. The first area Williams identifies is a broadly sociological approach. Williams contended that within this approach two primary traditions dominated the sociology of the media. The first tradition drew upon critical histories and histories of art. Williams notes that in more recent times such theories focused upon three areas; the social conditions of art, social material in art and social relations in art. The second tradition in the sociology of media is what Williams refers to as 'observational sociology' (ibid.: 16–20). Here, a further three sub-traditions are detected.

First is an area that is referred to as 'institutional approaches', which focused upon studied media institutions. Two broad perspectives have been of influence. Much initial work was performed in this area from a functionalist perspective to examine (often uncritically) the role of the media in market economics; it was characterised by a strong empirical methodology. A second perspective was conversely a critical Marxist approach that sought to identify the function played by the media institutions in the maintenance of capitalist society. Initial work by Wright Mills (1956) established the strong interrelationship of various élite groups in different areas of societal life. Schiller (1969, 1976) extended this argument into the ownership of the media.

The second area was influenced by the previously noted concern of social scientists with aspects of mass culture. This tradition has a strong ancestry within textual studies approaches, and Williams notes that in many ways these interests are similar to studies in art and literature. Early research in this area was concerned with the analysis of content and the 'selection and portrayal of certain figures' (Williams, 1981: 19). This early representational analysis was understood to be of considerable importance. It illustrated the links between systems of textual production, texts and cultural hegemony. Often referred to as 'neo-Marxist', the approach was primarily a study of text with little regard for how the information was received.

The third area Williams identifies is that of research concerned with the impact of the media. Williams notes how this tradition owes much to marketing and advertising, but also to audience research. Two strands are discernible: operational studies concerned with commercial research and critical research, and investigating the possible negative impact of a particular genre or media form. These early traditions were latterly incorporated into cultural studies, post-Marxist and more recently post-structuralist discourse theories.

Newbold et al. (2002: 40) note the continued significance of the tripartite division within media theory. 'In broad brush terms, the current complexity of the field can be accommodated within three of the most significant movements in media studies.'

Part I
Understanding the internet

2

Theories of technology and society

Introduction

Discussions of technology, particularly contemporary discourse surrounding the internet, are underpinned by deeply felt but often unarticulated assumptions of how technology and people interact. For the most part the relationship between technology and society, particularly new technologies, is assumed to be of a simple deterministic nature – the introduction of new technology causes social change. This belief lies beneath much (but not all) contemporary policy-making in the field of technology deployment: social problems can be ameliorated through the strategic use of technology. Indeed, this belief is voiced time and again with the emergence of new technologies and their supposed effects.

Historically, other versions of the relationship between technology and society have been proposed. In some instances these other 'readings' also contribute to contemporary discussions of how the internet is understood. In this chapter I want to examine a range of interpretations of the relationship between society and technology that inform contemporary discourse surrounding the internet. The focus here will be upon the 'deep' ideas that inform how we understand the internet. In examining this issue I argue that the way in which the internet is understood in much contemporary discourse, regardless of where that discourse is currently taking place, owes much to how technology and media have been understood historically in a number of Western societies. The particular experience of modernity and the development of technology in a number of Western countries resulted in a range of understandings of how technology, media and society interact. Through the various processes of globalisation, such as the dominance of Western intellectual traditions in higher education worldwide, these interpretations have become pre-eminent in how the relationship between society and technology is viewed.

Of course the argument that technology causes social change is a position in more established academic debate. The media theorist Bolter (2002: 77) argues that academic attempts to explain new media, often termed 'new media theory', can be divided into two broad camps: 'formalist' approaches – theories that 'appear to focus on "internal" or even "inherent" characteristics of the media' – and 'culturalist' approaches – theories that focus on 'characteristics that are "external"'. These broad positions are derived from an earlier debate between Marshal McLuhan and Raymond Williams concerning fundamental assumptions regarding the potency of technology and communication media.

The work of McLuhan played a significant part in the origins of media studies and mass communications, and informs a broad swathe of theories. McLuhan developed his ideas over a long period through a number of key texts: *The Gutenberg Galaxy* (1962), *Understanding Media* (1964), *The Medium is the Massage* (1967, with Quentin Fiore) and posthumously *The Global Village* (1989, with Bruce Powers). McLuhan is regarded as having raised many issues that concerned post-modern theorists in the 1980s (Ferguson, 1991). Briefly put, McLuhan sees human history as divided into four distinct, technologically oriented 'ages' or epochs: an oral/primitive age in which the dominant sense was aural, a literate age in which the visual sense became more important as visual artefacts rose in significance, a print age, during which the visual sense was dominant, and an electronic age, a multi-sensory period (McLuhan, 1962). The systems of media technology that dominated each age – auditory, textual, print and electronic – were the 'prime movers' in structuring human interaction and experience of the external world. Of the various ages, it was the 'print age' about which McLuhan had the greatest reservations. For McLuhan print technology caused human life to become fragmented: we were limited in our ability to use all our senses in communication, and thus the full range of human experience was denied to us. The electronic age offered salvation because it provided a new, more diverse and multi-sensory environment. Humans were able to express themselves and communicate far more fully in an 'electronic realm'.

McLuhan develops three key ideas from this broadly quasi-historical system. Firstly, remediation: the idea that all new forms of media borrow systems, techniques, styles and social significance from previous forms of media. Secondly, the extension of the sensorium: the notion that all technologies in some way seek to extend human capabilities and senses. The extension of senses radically changes experience of the world and, accordingly, culture. Thus developments in technology bring about changes in cultural form. Thirdly, the notion that 'the medium is the message'. McLuhan proposes that attention should be focused not upon the content of the media but upon the form in which it is delivered. It is the form of media rather than its specific

19

content that has the power to structure relations and human action. New forms of media thus bring about new forms of interpersonal interaction.

McLuhan's position can be contrasted with that of Raymond Williams, whose work later came to dominate the field of media studies and is regarded as being key in the field of (British) cultural studies (Turner, 1996). Although Williams's attacks on McLuhan's work led to a decline in McLuhanite studies, the advent of the internet has led to new interest in McLuhan's ideas. Williams's key work in this field, *Television: Technology and Cultural Form* (1974), is primarily sociological, in contrast to McLuhan's spiritual or psychological orientation. Deploying what became known as his 'cultural materialist' approach, he focuses attention upon the social conditions of technological and mediatic development and use. Three key aspects of Williams's work are of particular interest. Firstly, in opposition to the McLuhanite position that media technology changes mankind, Williams (ibid.: 129) proposes that technologies take forward existing practices: 'all technologies have been developed and improved to help with known human practices'. Secondly, technological development does not exist in a vacuum, rather it is tied to socially conceived goals – Williams proposes a 'social history' of technology as opposed to a purely technical account (ibid.: 14). Thirdly, the speed and direction of technological development are determined by the specific interest of certain groups: 'intention corresponds with the known or desired practices of a particular social group, and the pace and scale of development will be radically affected by that group's specific intentions and its relative strength' (ibid.: 129). Thus where McLuhan stresses the importance of technology in structuring human life, Williams proposes that nothing in a particular technology preordains its use or effects.

However, seeing media theory neatly divided into two broad camps, while affording considerable benefits in terms of simplicity, cannot account for the varied and wide-ranging field of new media studies. Indeed, Rice and Williams (1984: 55–6) argue:

> We should take advantage... of the new media to further specify and modify... theories... we may have to not only rethink current communication theories but, indeed, borrow from other disciplines and even construct new concepts and theories.

Furthermore, developments in the social theory of research have resulted in an increased degree of what is termed 'reflexivity' – the conscious and critical examination of one's own values and practices in research. Researchers are well aware of the criticisms mounted against the position to which they adhere, and modify their approaches accordingly. Thus the 'debate' between McLuhan and Williams did not stop with McLuhan and Williams; it has been conducted by proxy through numerous publications with increasing levels of complexity ever since (P. Jones, 1998). In addition to the research conducted in university departments it should also be recognised that much valuable work has taken place outside of academia that does not align itself with either a McLuhanite or a Williamsonian theoretical position (though as Bolter, 2002: 78 reports, popular theories often regard the formalist McLuhanite proposals of technological determinacy as given fact).

Therefore, while agreeing that the antecedent theories of McLuhan and Williams should be examined, I argue that a modified account of the underpinning theories of new media is needed. Instead of attempting to account for all approaches by drawing upon the work of two theorists, albeit highly influential ones, I propose that attention should be focused upon a wider field of analysis. We need to go beyond singularly academic debates concerning how we *should* understand the media and technology, and examine deeper, more cultural assumptions of how they *are* and have been understood. As Downing (1996) notes, the vast majority of research on the media has been conducted in the highly restricted laboratories of the USA and Western European countries. This is significant, for, as Hård and Jamison (1998: 1–5) propose, cultural tradition plays a considerable part in conceptions of technology. Our understanding of media, new media and particularly the internet emerge from a limited number of societies and reflect the values and understandings of technology in those societies. Likewise, academic studies and theories of media draw upon very particular conceptions of technology and its relationship with society. Such theories are not 'universal' but come from a particular historical tradition.

In order to understand the conceptions of technology and in particular the internet that are predominant, we should look at broader and deeper conceptions of how technology works. More specifically, attention should focus upon notions of technology and its relation to society within the social and cultural forms of modernity. In this and later chapters I explore a number of these underlying assumptions, and argue that we should recognise the historical and political 'situatedness' of how the internet has been understood. In the remainder of this chapter I examine how technology has been thought of, and how the way in which technology intersects with society has been understood from three broad perspectives.

Technology

Understanding what is meant when we refer to technology is problematic, as many different interpretations operate simultaneously. In the most reportive (how it is used) and prescriptive (the ordinarily used interpretation that appears in dictionaries and normal parlance – Dusek, 2006: 28) sense, technology is simply the examination of mechanical and scientific practices: 'the study or use of the mechanical arts and applied sciences' (Allen, 1990: 1252). However, this definition does not sufficiently describe how the *concept* of technology is understood. Outside of the purely semantic meaning of the word are numerous other ways of understanding technology as a concept or class of objects. Kline (1985) cites three further interpretations of technology in addition to the one noted above, and argues that technology should be understood in totality with the social systems that make it possible and make use of it. While useful, this definition is perhaps not specific enough to identify exactly what we mean by technology and what we do not.

An alternative definition of technology, one that emerges from anthropology, conceptualises technology as that aspect of human behaviour that distinguishes humans from other forms of life. This is the argument put forward by Chase (1929: 23), who, paraphrasing Thomas Carlyle (who was in turn paraphrasing Benjamin Franklin), states: 'all authorities agree that man is the tool-using animal. It sets him off from the rest of the animal kingdom as drastically as does speech.' Similarly Noiré, a philosopher of language, notes that the human is a 'tool-making animal' (quoted in Gehlen, 1983: 205). Three points may be made in relation to this view. First, technology extends and corrects human deficiency. Second, technology is conflated with any form of artifice.

Third, it is the use of technology that separates humanity, regarded as *homo habilis*, from other animals, an argument extended in the work of Postman (1993), who regards language as a technology.

Yet a further definition seeks to link technology to its etymological root, *techne*, meaning art or craft. Miller (2003: 1) contends that the term technology emerged first at the turn of the seventeenth century, referring to 'speaking or reasoning about the arts'. Its meaning gradually changed until 'technology came to mean the application of scientific principles to the useful arts, or rather a new way of making and doing whose basis is experimental or theoretical science' (ibid.).

This technologically optimistic description is problematised by Winston's (1998) historical study of communications technology. Winston makes use of the anti-science approach of Kuhn (1962), who proposed that science was not the steady, progressive, incremental accumulation of knowledge but instead a continuous series of stable periods interrupted by bouts of revolutionary discoveries. In Kuhn's model, science is not a continuing and incremental body of knowledge; rather, it operates in terms of 'paradigms', accepted bodies of knowledge and explanations. Over time, criticisms build up against a particular body of knowledge until eventually a scientific revolution takes place in which a new theoretical interpretation is proposed and becomes widely accepted. Kuhn's work challenges the essential progressive notion that science leads towards an eventual truth – 'We may have to *relinquish* the notion, explicit or implicit, that changes of paradigm carry scientists and those who learn from them closer and closer to the truth' (ibid.: 170; emphasis in original). Instead, we should acknowledge that all knowledge, including scientific knowledge, is relative.

Winston (1998: 3) continues this argument and asserts that technology is the performance of historically situated scientific principles:

> communication technologies are... performances... of a sort of *scientific competence*. Technology can be seen as standing in a structural relationship to science. Technologies are, as it were, utterances of a scientific language, performances of a scientific competence. (Emphasis in original.)

In this interpretation technology is not forms of action, rather it is deeply tied to scientific principles. Furthermore, the definition assumes a socio-historical position wherein technology is understood as the expression of *current* scientific paradigms as opposed to the expression of universal truths.

The relationship between technology and society within modern discourse

As technology may be understood in different ways, so may the relationship technology has with society. In proposing a philosophical-anthropological orientation to the study of technology, Feenberg (1999a: 1–17) contends that within modern discourse the relationship of technology with society has been conceived of in a number of different ways.

Like Winston, Feenberg is implicitly working within a Kuhnian theoretical framework. Feenberg proposes an alternative system of understanding of technology within which it is perceived as neither the expression of universal truths nor the result of linear progress.[1] Feenberg argues instead that, like scientific paradigms, discourses of technological understanding emerge from 'local' historical conceptions and are interwoven with political and social projects. He asserts that the development of such an analysis is key to grasping a sense of technology's significance, asking 'how can one study specific technologies without a theory of the larger society in which they develop?' (Feenberg, 2004: 73). The categories Feenberg identifies offer a richer and more detailed account of the beliefs underpinning accounts of technology, and consequently new media, within modern and late-modern societies. It is to an examination of these categories that this chapter now turns.

Instrumentalism

The instrumental understanding of technology holds two beliefs at core. First, technology is 'essentially' neutral – it carries no values; and second, technology is subservient – it does what we want it to. Technological artefacts are inherently different from cultural artefacts: they are purely means-oriented, they are designed to complete tasks, to allow people to achieve a goal. Conversely, cultural artefacts such as works of art seek to communicate and carry the values of a society.

This interpretation of technology, that it is distinct from other forms of manufacture, emerged due to the exclusion by the classical humanities disciplines of science and technology from their field of study (Feenberg, 1999a: 1; Escobar, 1994). This approach or attitude is perhaps best exemplified by the British essayist Alexander Pope's famous assertion regarding what should be studied: 'the proper study of Mankind is Man' (1870: 225–6). Pope was demonstrating a deep-seated disdain for science and, by association, technology that was widespread among the European intelligentsia. It is only with the gradual transformation of cultural activity and education that occurred with the Enlightenment and the advent of modernity in certain European countries that accounts of technology became more central. However, as Winner (1987) notes, these accounts tended to regard technology in a purely functional manner. Technology was rarely foregrounded in accounts – it was a device to achieve economic or political goals. This account of technology persists today, for, as Winner (ibid.: 2) states, there is still an open 'tendency... to see the matter solely in terms of economics and economic history'. In addition to the relegation of technology to a position subordinate to economics or politics, instrumental accounts of technology tend to centre on certain unchallenged assumptions. Winner (ibid.: 25–7) states that:

> In the conventional perspective works of technology are more than certain: they are doubly certain. Since human beings are both the designers and makers of their creations... they know exactly how things are put together and how they can be taken apart... that which men have made they also control. This is common sense... technical means are by their nature mere tools subject to the will of whomever employs them... Technology is essentially neutral. In the conventional way of thinking, the moral context appropriate to technical matters is clear. Technology is nothing more than a tool.

Furthermore, technology 'appears as purely instrumental, as value free. It does not respond to inherent purposes, but is merely a means of serving subjective goals we choose as we wish' (Feenberg, 2003: 3).

Instrumentalist readings of technology still circulate widely today and are evident in many accounts of online activity. Accounts that study how social categories such as race, gender, age and class influence access are implicitly instrumentalist – the social categories are conceptualised as existing prior to their manifestation online (Surry and Farquhar, 1997). Thus much progressive examination of the lack of engagement of women, ethnic minorities and the low waged in new media, of the problems of the digital divide and other issues of access makes use of an instrumental conception of technology. Similarly, fields of practice such as community informatics that actively use the internet and information and communication technology to further community needs incorporate an instrumentalist stance.

Thus instrumentalism as an approach to technology should not be regarded as a passive, historical, default position held by those who do not engage with more critical readings, which would seem to be the argument of Feenberg (2003) and Escobar (1994). It is a lively and defiantly human-centred approach that asserts human action as the sole motivating factor behind the effects, impacts and consequences of technology. Instrumentalism identifies human agency and social factors as the 'prime movers' in social life and rejects claims of helplessness in the face of technological change.

Determinism

Along with the instrumental reading of technology, a second and persistent understanding of technology is evident within the discourses of modernity. Broadly referred to as 'determinist', this category has as its mainstay a belief in the potential of technology to bring about social change on a macro or societal level. Technological determinism has proven a strong and persistent strand of thought in understanding the role of technology within modern Western thought, even though it seems rarely explicitly stated. Marx and Smith (1996: ix–xv) contend:

> A sense of technology's power as a crucial agent of change has a prominent place in the culture of modernity. It belongs to the body of widely shared tacit knowledge that is more likely to be acquired by direct experience than by the transmittal of explicit ideas.

Similarly, Bimber (1996: 80) proposes: 'Technological determinism seems to lurk in the shadows of many explanations of the role of technology.' With regards to a general description of technological determinism, Heilbroner (1996: 69) summarises the argument as follows: 'Machines make history by changing the material conditions of human existence. It is largely machines… that define what it is to live in a certain epoch.'

Feenberg (2003: 1–2) contends that such a trend emerged out of notions of progressivism within the Enlightenment, and more specifically an engagement with the progressivism of Marx[2] and even Darwin. In post-Enlightenment European society, progress came to be broadly equated with an acknowledgement of technology's power; 'progressivism had become technological determinism' (ibid.). This form of understanding has proven highly persistent and popular. It continues to manifest itself in numerous formats. For example, one particular and contemporary understanding equates the deployment of technology with improved social conditions. A number of populist accounts regard the deployment of digital technology as a necessary precursor to the development of a 'knowledge economy' or 'information society' (Lyon, 1988). In a much-cited argument Toffler (1971: 25) lists numerous instances where technological innovation has resulted in improved social conditions and states: 'behind such prodigious economic facts lies that great, growling engine of change – technology'. Numerous national governments have sought to deploy technology rapidly in pursuit of economic and social development. For example, in a statement by the Welsh Assembly Government (undated) detailing its 'information and communication strategy' it is contended that:

> Many of us are now using computers, mobile phones and the internet... These technologies have the potential to transform society and the economy in Wales; they are already doing so in many parts of the world. The choices we make now – about which new technologies we use and, more importantly, how we use them – are crucial to the future of Wales and will help us to create a Better Wales!

Similarly, the Malaysian government has instigated, and to a degree acted upon, plans to 'leapfrog into the Information Age', developing a 'multimedia super corridor', a region of technological development incorporating purpose-built cities and a university all underpinned by highly developed technological infrastructure (Multimedia Development Corporation, 2000).

The nature of the transformational power or agency of technology is not, however, consistent in all understandings of technological determinism. Marx and Smith (1996: xii) contend: 'The idea of technological determinism takes several forms, which can be described as occupying places along a spectrum between "hard" and "soft" extremes.' Accounts such as those noted above of the power of technology manifestly to change, alter or improve a society, usually thought for the better, may be regarded as a particularly 'hard' version of the technological determinist thesis. Specific technologies are understood to result in definite changes in society. 'Softer' versions of technological determinacy may instead integrate technology into a basket of changes taking place in a society.[3] 'Softer' versions of technological determinacy may appear similar to the instrumental understanding of technology, where technology is seemingly demoted to a position in which its change-causing potential is downgraded to the degree that technology is just another aspect of social life. However, such a position is distinct from instrumentalism, in that technology may still have some effect, whereas in instrumentalism technology is simply a passive background component of the world and not a foregrounded active aspect. Furthermore, technological determinism provides an epistemological viewpoint from which to examine notions of technology; Heilbroner (1996: 77) contends that 'Technological determinism gives us a framework of explication that ties together the background forces of our civilisation, in which technology looms as an immense presence, with the foreground problem of the continuously evolving social order in which we live.'

Bimber (1996: 83) challenges this reading, and instead proposes that rather than using notions of 'hard' and 'soft' determinacy, strict technological determinacy is expressed only by those adhering to what is termed a nomological version. Such an approach is characterised by a reference to ideas of science and nature as orientating the direction of development and an utter rejection of the social in accounts of technology's potency. This reading incorporates two claims: that 'technological developments occur according to some given logic, which is not culturally or socially determined, and that these developments force social adaptation and change' (ibid.: 84). Such dual claims seem to capture the essence of technological determinism: that development occurs in line with a progressive truth-revealing logic and not a social-determined one.

Substantivism

A range of attitudes broadly termed substantivist challenge the determinist belief in the neutrality and truth-revealing nature of technology (Feenberg, 2003: 2). As instrumentalism and determinism are understood to have emerged from empiricist and positivist tendencies within Enlightenment thought, substantivism is understood to have arisen from the distrust of technology (ibid.: 2–3) and the reassertion of the 'natural' found within Romanticist discourse.[4] Similar to determinist discourse, substantivists contend that technology can directly intersect with and modify social life. However, substantivism avoids the utopian and optimistic tendencies that characterise determinist accounts, and instead maintains deep reservations about technology. Technology is understood inherently to subjugate the user to systems not initially declared in the operation of technology. Such a belief reaches its most eloquent form in Heidegger's (1977) *The Question Concerning Technology*. Heidegger proposes that technology is far from the neutral or simply goal-oriented system determinists or instrumentalists would claim. Rather:

> we are delivered over to it in the worst possible way when we regard it as something neutral; for this

conception of it, to which today we particularly like to do homage, makes us utterly blind to the essence of technology. (Ibid.: 42)

Technology contains an 'essence': Heidegger envisages that technology is not about achieving goals but about 'revealing' or bringing forth the use of a resource. However, 'modern technology' is fundamentally different from what Heidegger regards as ancient technology. The form of revealing is primarily different because of the physics-based nature of modern technology that allows for the ordering of a 'standing reserve' at the behest of humans. This is opposed to the fundamental primacy of natural forces in old technology. However, Heidegger regards modern technology as inherently insidious, as humans do not control technology; humans form part of the system of standing reserve. Humans are 'enframed' by technology and technological systems and lose their freedom through their incorporation into technological systems. While Heidegger offers a radical reading of technology, it lacks a sociologically 'critical' aspect in that 'fault' is understood to lie with modern technology as an entity as opposed to the more critical conflicts and power relations that underpin modernity.

Latour (1988) develops the idea of the potential of technology to modify life in a different direction. Noting that in technology's fashioning certain potentially political possibilities of use are incorporated, Latour, through the micro-examination of a simple piece of technology, a door closer, examines the way in which technology can impact upon daily life outside of macro-political issues. Such an approach is interesting, as it seeks to explore the ways in which technology intersects with human life outside of the macro-explanatory models used in social and philosophical studies of technology (particularly those prevalent in deterministic thought). In concurrence with actor-network theory,[5] Latour proposes that through the use of technology we integrate ourselves into 'networks' where both technological and human actors must be considered as having agency (effective if not volitional). Latour examines how we delegate tasks to technology, the labour-saving device, and how we are in turn regulated by the technology. Technology, argues Latour, regulates our action through our integration with technological systems. We 'instruct' technology to perform the tasks that are too mundane or difficult for normal people. However, in 'delegating' to technology we accept the way in which the technology will perform the task – that the task will be performed in a specific fashion. This fashion, way of performance or inscription then sets the way in which the task is carried out, and coincidentally encodes the ideological position of the technology's manufacture into the task, and we work with this way of doing the task. The possibilities of use in turn fashion the action around technology.

Substantivist thought also incorporates a spectrum of opinion that links the subjectifying nature of technology with specific political projects, a radicalising and politicising of Romantic thought. Technology is conceptualised as inherently political. Winner (1996: 28) contends:

> At issue is the claim that machines, structures, and systems of modern material culture can be accurately judged not only for their contributions of efficiency and productivity, not merely for their positive and negative side effects, but also for the ways in which they can embody specific forms of power and authority.

At the core of such claims lies a different conceptualisation of the nature and understanding of the origin of technological artefacts to that of the instrumentalists and the determinists. Here, in an engagement with Kuhnian theory, substantivism explicitly challenges the notion that technology is a truth-revealing (or revealed) phenomenon; contrarily, technology arises from, and is broadly shaped by, society (Freeman, 1992).[6] Substantivists argue that in terms of production, technology cannot be distinctly discerned from other forms of cultural production and, as with all forms of cultural production, technology is inherently stained by the situation of its material and economic production. Substantivism offers a theory not only of the effects of technology upon society but also of the effects of society upon technology. Qvortrup (1984: 7) proposes that new technology 'cannot be properly understood if we persist in treating technology and society as two independent entities'. Technology needs to be understood as a component of society. Consequently, and most importantly, technology is in essence determined by the society in which it originates. It is an artefact of a civilisation and not a progressive quest towards truth. Technology is not the neutral artefact presumed by instrumentalists and determinists. For a substantivist, technology is inherently compromised by its site of production. Marcuse (1968: 224) proposes:

> Specific purposes and interests of domination are not foisted upon technology 'subsequently' and from the outside; they enter the very construction of the technical apparatus. Technology is always a historical-social *project*: in it is projected what a society and its ruling interests intend to do with men and things. Such a 'purpose' of domination is 'substantive' and to this extent belongs to the very form of technical reason. (Emphasis in original.)

It is this critical and dystopian dimension, that technology contains the insidious 'will' of its situation of manufacture, that distinguishes substantivist accounts from the utopian progressive accounts of technology proffered by determinists. Technology is inherently a problematic system of control for substantivists, a form of instantiated power.

Contrasted with the instrumental and determinist interpretations of technology, substantivism offers a highly pessimistic and critical reading of the further integration of social functions within systems of technology.

Conclusion

The view presented here, of the existence of differing interpretations of technology's interaction with society, is of considerable importance. As will be shown in later chapters, the cultural base of beliefs surrounding the internet is rarely acknowledged. However, contemporary discourse concerning the internet is underpinned by historic and particular beliefs; the internet is understood in a fashion that owes much to existing interpretations of technology, the media and politics in Western societies. It will be argued in the following chapters that the internet is widely understood to possess a range of characteristics that are not present in older forms of media. Furthermore, such characteristics afford a means by which forms of political action may take place. The politics assumed to be enabled by the internet has proven doubly interesting to scholars. First, it seems to instantiate new opportunities or spaces for public and critical discourse, a feature understood as absent within older media. Second, such politics is understood to be of a radical nature, expressing opinions rarely found within mainstream mass media. Such readings draw from the three traditions summarised above. Technology is regarded as effective and deterministic to society, and even individuals, as it may allow for the deployment of contrary opinion through various forms of content production and dissemination. It is also regarded in an instrumental fashion, subject to differing political agendas. However, at the same time it also entails a certain substantively democratic potential – the use of the internet is understood to encourage democratic action (while a substantive viewpoint is that technology is understood in terms of politics being encoded within a technology, this version regards the technology as essentially progressive).

Identifying differing interpretations of the interaction of technology and society and the conflation of these interpretations in accounts of the internet is significant for a number of reasons. Firstly, while differing opinions did, and still do, exist in the conceptualisation of technology, it may be problematic to group them into 'schools of thought' as has been done here. However, it should be noted that the collapsing of often-contradictory traditions into broader fields of argument has been noted as a feature of modern discourse. The boundaries that divided fields of knowledge in preceding eras are understood to break down in conditions of late- or post-modernity. Indeed, Berman (1982: 17–22) argues that late- or post-modernity is characterised by the collapsing of previously stable divisions of knowledge and the appearance of new 'fissures'. Giddens (1994: 184) speaks of 'scenario-thinking'[7] the ways in which we reconceptualise the world according to different agendas brought to prominence by the changes in our ontological frameworks. Our experiences of late-modernity result in our old categories of thought being merged and blurred. This is not to say that the previous divisions no longer have relevance, despite the difficulty in discerning them – they often possess such cultural inertia that they continue to inflect discourse without being overtly visible. As to an understanding of the internet, culturally we might know something but be unaware of the political orientation and underpinning of such knowledge.

Secondly, once it is accepted that numerous principles underpin conceptions of technology then no single reading of a technology or its effects should be adopted. If multiple and equally valid readings of technology are possible then the absolute fixed interpretation of a technology is challenged. Instead, a reading of a technology must be regarded as a field of discourse or a site of contention. Therefore, rather than studying the actual technology of the internet, attention should be paid to the social discourses surrounding it. The internet should be regarded as an aspect of social life, narrated through discourse and conceptions of the world.

Recognising that technologies such as the internet are known only through our narrations is highly significant, and it is a description of the internet in contemporary journalistic, humanities and social scientific literature to which I will now turn.

Notes

1   Broadly understood under the rubric of 'critical theory', Feenberg's new approach to technology advocates the position that while technology entails specific social values, such values do not explicitly close off counter-appropriations or forms of use.

2   A general predilection to technical determinist theory within Marx's work is not in dispute. Marx (1955: 95) contends that 'in acquiring new productive forces men change their mode of production, and in changing their mode of production they

change their way of living – they change all their social relations. The hand mill gives you society with the feudal lord; the steam mill society with the industrial capitalist.' However, the details and conditions of technological development within Marx's thought remain contentious. Elster (1985: 143) proposes that 'Marx believed, paradoxically, both that technical change was the central fact in all world history and that it was a phenomenon uniquely characterising capitalism.' The use of a technology is specifically tied to material relations – relations that can only exist within a capitalist framework. However, this seems contradictory to the central nature of 'history' that is essentially progressive. Elster contends that while such a contradiction is a problem, it is not insurmountable, as any theory seeking to offer an overarching theory of social change will contain contradictions and problems.

3   They may even question the reason for the emergence of technologies within specific societies. Of particular note are the myriad explanations offered for the increased pace of innovation that occurred in certain European countries from the middle of the eighteenth century onwards. Regarding the plethora of explanations, ideological and religious predilections to certain work patterns and the development of certain forms of fiscal exchange, Marx and Smith (1996: xiii) propose that 'almost every identifiable attribute of early western societies has been proposed as the putatively critical factor'.

4   The focus here upon certain elements of Romantic discourse perhaps ignores more substantial components of the Romantic movement and does it an injustice. Poulet (1966: 40) proffers that at core a Romantic is someone who 'discovers himself as centre'. This initial disregard of the unknowable or unreachable world of objects in favour of the 'I as subject' constitutes a more significant critique of empirical knowledge than a seeming distrust for technology.

5   The actor-network theory developed out of work by Callon (1991) within the social study of science and technology. It sought to understand the interaction of individuals within environments. A specific methodology and range of terms were

developed to explore the way in which people interact with their immediate surroundings. Action is specifically regarded as being mediated by both other humans (actors) and material and social artefacts (actants). Understood as a collective, a social and material environment composed of actors and actants is understood as a 'network'.

6   Interestingly, Freeman attempts to deploy paradigmatic theory within a determinist methodology, regarding determinism as flexible enough a system to accommodate a variety of theories to explain the 'driving force' of science and technological development. Determinist theory usually concurs with progressive accounts of technological change as either a goal-oriented direction or an accumulation of minor changes and improvements upon technology. In both accounts, technological change is progressive. Freeman's paradigmatic account, however, proposes that scientific change may occur in other fashions, though this makes little difference to determinist arguments.

7   Giddens is referring to the emergence of a risk society out of a societal form in which notions of chance and fate had been devolved to unearthly and unknowable entities. In the risk society, chance becomes measurable and economies emerge to codify and possibly indemnify the individual from uncertainty. However, greater consideration is also accorded to instances of manufactured risk.

3

The idea of the internet

Introduction

This chapter examines the way in which the internet has been discussed, written about and generally understood in a range of popular and academic texts. As noted in Chapter 1, one of the aims of this book is to explore the way(s) in which the internet has been popularly understood. In Chapter 2 I argued that there were certain beliefs or ideas about the relationship between technology and society seated deep within contemporary culture. These ideas have long histories within Western intellectual traditions and continue to 'inflect' how new media such as the internet are understood and thought about. This chapter explores a different facet of how we think about the internet. The intention here is to look at popular ideas concerning the qualities or characteristics we think the internet possesses, and how these distinguish the internet from preceding 'old media'.

This approach is markedly different from a scientific or technically orientated description of the internet. In this chapter I am not seeking to examine what the internet 'really' is, to reveal new knowledge about its impact or to make new pronouncements about its revolutionary potential – these are well-worn topics of discussion and are covered more than adequately elsewhere (for an excellent technically orientated history of the internet see Abbate, 2000). Instead I wish to explore just what is it we are saying about the internet. What is it we believe the internet can actually do? What qualities do we believe new media possess that older media do not? In short, what is new about new media?

In exploring these issues I want to steer away from a 'revelatory' approach to knowledge of the internet and towards an appreciation of how it has been 'represented', understood and discussed in contemporary culture. This approach is part of a broader form of social enquiry, anthropological in origin, that seeks to examine the interpretations people make that allow them to live and experience the world in a meaningful manner. As Giddens (1976: 160) argues, much sociological enquiry is about exploring, charting and examining the knowledge that is 'constituted or produced by the active doings of subjects'.

Thus, the intention is not to provide a factual or technical description of the internet, but to examine the way in which it is popularly understood, thought about or interpreted. Being an account of the popular description of the internet, the ideas covered here are not buried or in some way hidden from view, awaiting discovery. Quite the opposite: I am seeking to describe the popular, everyday understanding. These views are everywhere and all around us. They exist in popular and academic texts, and the goal here is to chart how the internet is thought of in popular discourse, illustrated through contemporary journalism, popular discussion and academic discourse – specifically texts concerned with the internet from the social sciences.

It is also worth noting that viewing popular and academic interpretations of a technology (or any social phenomenon for that matter) together is not an outlandish thing to do. The popular understanding of a technology owes much to how it has been understood in academic literature. Indeed, popular and academic interpretations of technology are deeply linked. Academic knowledge certainly informs popular culture, and even more certainly is influenced by popular culture (Allender-Hagedorn and Ruggiero, 2005). Science and technology studies have examined the way in which scientific thought influences and is influenced by popular culture. A similar process occurs in the social sciences and humanities. Terms and concepts developed in academic settings have a way of permeating into popular culture. Giddens (1991: 14) notes how the integration of texts produced within academic or professional contexts into contemporary cultural life is a feature of modernity, arguing 'such writings are part of the *reflexivity* of modernity: they serve routinely to organise, and alter, the aspects of social life they report on or analyse' (emphasis in original). Academic and technical texts often set the general orientation and direction of understanding. Giddens (ibid.) asserts that 'such knowledge is not incidental to what is actually going on, but constitutive of it'.

In this chapter I argue that there is a distinct 'description' or understanding of the internet in operation in popular culture, and that this interpretation is constituted and articulated through and by both popular and academic texts. Dawkins (1989: 192) notes that particularly successful interpretations, fashions or ideas have a way of spreading and self-propagating, becoming the common understanding, and refers to such units of cultural knowledge as 'memes'.[1] A key facet of the description of the internet examined here is the meme or publicly accepted idea that the internet is a media form that possesses certain unique qualities that distinguish it from previous forms of media communication technology.

The internet: different from other media

Before moving on to this current description of the internet, in which it is viewed in some way as different from older media (and therefore worthy of study), it is worth noting that this has not always been the case. The early study of new media from a broadly humanities and social science approach produced a number of accounts that considered the nascent internet and other forms of computer-mediated communication solely as technologies of 'interpersonal communication', the study of which should be 'relegated to the domain of other fields' (Morris and Ogan, 1997: 2). New media technologies were considered similar to the telephone – a technology of one-to-one communication – and similarly for the most part ignored by mass-communications scholars. For example, DeFleur and Ball-Rokeach (1989: 335–6) argued that 'even if computer literacy were to become universal, and even if every household had a personal computer equipped with a modem, it is difficult to see how a new system of mass communication could develop from this base'. This approach ran counter to the academic study of mass media in which there was (and still is) a distinct tradition of studying the media used to convey messages. Mass-media scholars have traditionally been concerned with issues of media content, the effect of media upon audiences and the institutional systems of ownership, management and distribution. In a relatively early paper addressing the issue of whether new media constituted a technology of mass communication or interpersonal communication, Rice and Williams (1984) argued that new media technology spanned a traditional divide in the categorisation of communication technology into systems of interpersonal and mass communication. New media should be considered *both* a means of interpersonal one-to-one communication and a means of mass communication. This was possible because of the particular characteristics or qualities that the media possessed. In this interpretation new media, of which the nascent internet was but one, are certainly worthy of study as they are understood to possess certain characteristics that distinguish them from preceding media.

The issue of whether the framework of analysis that has been used to examine mass media can be used for the internet is of some contention. For the most part media scholars assert that while the internet may be different in many ways, the methods that served so well to understand existing media work just as well to understand new media. However, an increasing number argue that the internet is just too different from old media and new methods must be developed (Lister et al., 2002; Dovey, 2002; Merrin, 2008).

Web 2.0

This argument has been strengthened considerably with the growing interest in what is termed Web 2.0, a suite of applications, practices and user behaviours which are (or are considered to be) more user-centric than preceding internet applications. Web 2.0 is a recent term, and refers to a number of developments in media content production, dissemination and use. Many sources credit Tim O'Reilly, vice president of O'Reilly Media, and Dale Dougherty, an internet entrepreneur and previously an academic, with inventing the term – the first public use being the name for a conference in 2004. It has been described as a range of technologies; a suite of applications; a business model; a form of civic empowerment; and even a 'mind-set'.

However, many of the applications and practices considered part of Web 2.0 had been in existence for some time. Web logs or 'blogs', for example, existed in an ancestral form as a newsgroup, mod.ber, from as early as 1983. The term blog was coined in 1997 (Wortham, 2007). Furthermore, a number of authors argue that there is nothing technically new in Web 2.0 applications; they simply do what the internet was capable of before due to the specific characteristics of internet technology. Perhaps the best way to conceptualise Web 2.0 is to regard it as a milestone in the public interpretation of the internet. Web 2.0 is a means by which the meme of the internet as different from other forms of media is made more understandable.

The characteristics of the internet

Listing the various qualities that are understood to set the internet apart from other forms of media has been attempted on a number of occasions. Interestingly, many authors consciously conflate what are proposed to be actual qualities of new media with the use or consequences of new media. This is particularly interesting as it identifies the importance of technology not as what it *is*, the actual characteristics of technology, but as how it is used or the consequences of its use. Additionally, there is a merging of the infrastructural aspects of the internet – the computers, networks and technical protocols that make the transmission of data possible – and the software applications that make those transferred data suitable for consumption by non-specialist users. Some of the characteristics listed in the next paragraph are manifestly qualities of the 'hard' infrastructural aspects, while others really only emerge in specific usages of particular software applications and only occur under specific conditions, a point that will be elaborated upon in later chapters. Five texts have been selected for their significance and general representiveness of attitudes towards the internet. Three are distinctly academic, one of which was written for a popular audience. Of the final two, one is a cross-over from academia to web design and the last is from an influential new media technical publisher.

Within academia there have been numerous accounts of the qualities or characteristics of new media. McQuail, formerly professor of communications at the University of Amsterdam in the Netherlands and visiting professor at Southampton University in the UK, offered an early summary of these qualities that, in many ways, set the tone for later accounts. McQuail's (1986: 8) work drew upon a range of predominantly optimistic texts about the (then) predicted digital revolution, and argued that new media were characterised by their 'abundance of production and supply, freedom of choice, interactivity, loss of central control, decentralisation, search and consultation'. Similarly, van Dijk (2006: 6–9), professor of communication studies at the University of Twente in the Netherlands, identifies three different characteristics: convergence – the integration of telecommunications, data communications and mass communications; interactivity – action and reaction between user and text; and digitisation – the transference of all forms of content into a format that can be transmitted between computers over data networks, manipulated and reproduced. A far longer list is offered by Logan (2007), emeritus professor at the University of Toronto's Physics Department, who lists 14 characteristics that together define new media: two-way communication; ease of access to and dissemination of information; continuous learning; alignment and integration; community;

portability and time flexibility offering freedom over space and time; convergence of different media; interoperability; aggregation of content; variety; the closing of the gap between producers and consumers of media; social collectivity and cooperation; remix culture; and the transition from products to services. Negroponte, founder and then head of the Media Lab at Massachusetts Institute of Technology, provided a very influential and popular account aimed at a wider audience. Negroponte (1996: 229) argues that digital technology has four qualities 'that will result in its triumph: decentralising, globalising, harmonising, and empowering'. In a text targeted at professionals in the field of web content production, Manovich (2001: 18–45), a professor in the Visual Arts Department, University of California, San Diego, lists five characteristics of new media: numerical representation, modularity, automation, variability and transcoding potential. Manovich (ibid.: 49–55) also identifies a number of misnomers in contemporary discussion surrounding new media. He contends that the much-cited quality of interactivity and the requirement to be digital are not essential descriptors of new media. Finally, on a website dedicated to the collaborative production of documentation for open-source software, the technical publisher and new media innovations company O'Reilly argues that content (specifically artistic and creative practice) created and delivered over new media has seven characteristics: new media art is digitised, malleable, convivial, open, topical, applied and constrained (Anonymous, 2007).

These descriptions point to a general view that new media are regarded as possessing a set of qualities quite unknown or at least unrecognised in older systems of media. Of the five accounts mentioned, typical of numerous texts in the field, McQuail offers a particularly useful analysis (one that is continued by van Dijk). In seeking to examine the broad conception of the internet, it may be useful to examine his ideas further. McQuail's conception draws on theories proposed by Bordewijk and van Kaam (1986) concerning the flow of information between individuals and media institutions in what they term 'information societies'. Bordewijk and van Kaam developed a model that plotted the movement of information between 'information service providers' and 'information service consumers' (ibid.: 16). McQuail (1986, 2001) develops this model and makes effective use of it in a number of papers. The model provides a way of categorising forms of media by describing the way in which information is stored and its dissemination controlled. Two axes are used. The first axis describes the storage of information and two possibilities are identified: information is either stored centrally in a distant location or it is stored by individuals. The second axis concerns the control over the access to the information – is control held by a central agent or by individuals? Using these two bisecting axes, four possible positions are then possible: allocution, where centrally stored information has its distribution controlled by a central agent; consultation, where centrally stored information is consulted by users at the users' discretion; registration, where access to individually stored information is controlled centrally; and conversation, where individually stored information is used at the behest of individuals.

McQuail argues that, for the most part, old media can be understood as operating in a predominantly allocutive sense (centrally stored information disseminated at centrally controlled times, such as television broadcasts), but also to a lesser degree in a consultative sense (centrally stored or produced – archives of texts, for example). In addition to these two patterns, McQuail argues that new media are best situated within the other two cells of the model, registrational and conversational. New media and in particular the internet are considered potent, in that they shift the balance of power between producers and consumers of media. The remainder of this chapter will be concerned with mapping and describing the qualities of the internet that make this shift possible. These perceived qualities are grouped into four main fields: human-technology interactivity, interpersonal communication, the production and dissemination of content by users and the individualisation of content.

Human-technology interactivity

Despite the above-noted protestations of Manovich (2001), interactivity is a quality widely assumed to be present in new media and is perhaps understood to be *the* defining quality – Rogers and Chaffee (1983: 25) proposed that 'scholars are going to have to shift toward models that accommodate the interactivity of most of the new communications technologies'. Interactivity is widely understood as *the* characteristic or quality that demarcates new media from old. Such a view is not restricted to the academic community – the innumerable benefits of interactivity have been advocated and well documented by (and to) the business and arts community. However, as Jensen (1998) notes, interactivity is far from a being a fixed concept: its use may be very much dependent upon context and interactivity has been understood in different ways by different disciplines. Jensen lists several academic subjects in which interactivity plays a part. He argues that the current dominant understanding of interactivity in relation to new media owes much to how interactivity has been understood in sociology – being the links between communication and interpersonal interaction; communication studies – more correctly cultural or literary studies concerning post-modern theories of the changing nature of the text; and informatics – specifically the field of human-computer interaction.

The interactivity that is attributed to the internet is a combination of these three traditions. In this conception technologies such as the internet allow for systems of control over the flow of information and over the selection of content presented by the media (Chapman and Chapman, 2000). This quality is regarded as of considerable significance to researchers in the field, but there is some disagreement concerning its exact nature. Downes and McMillan (2000: 158) identify two distinct traditions shaping research in this area. The first focuses upon 'interaction in human communication' and is primarily concerned with interpersonal communication. This is a set of ideas that will be addressed in more detail below. The second tradition concerns the interaction of humans with machines, and manifests itself most commonly within the academic discipline of human-computer interaction (HCI).

The HCI discipline emerged primarily from computer and information science. Shneiderman (1997) and Crawford (2002) both assert that the discipline is usually focused upon computers as a distinct form of technology from other forms of communications technology. In many accounts, interactivity is regarded as responsiveness on the part of technology to an action by the user. In an introductory text Gorard and Selwyn (2001: 52) describe interactivity as follows: 'At a broad level, interactivity... gives the user a degree of influence and control in their use of the program.' From this perspective, interactivity is equated with the notion that some human action or method of input affects the future presentation of information. Interactivity is understood as the ability of the media technology to engage in a form of 'communication' with the user; over a series of exchanges the technology modifies its state in direct response to some form of prior input from the user. Rogers (1986: 5) refers to this quality as 'talk back'. This phrase refers to the idea that control of timing, message content, the sequence of the communication act and alternative choices are implicit and fundamental aspects of interactive media technologies (Williams et al., 1988). In the first of a series of texts, Crawford (1990: 104) proposes that interactivity should be regarded as a 'circuit through which the user and computer are apparently in continuous communication'. In a later text the same author proposes that interactivity is fundamentally 'a cyclic process in which two actors alternatively listen, think and speak' (Crawford, 2002: 3). The terms 'actor', 'listen', 'think' and 'speak' are used metaphorically to explain that the interaction between user and computer is akin to an interaction between two people. Interactivity is understood by Crawford as the propensity of an 'actor' to respond to previous actions of the other 'actor'. For example, a website changes the displayed content according to user action, responding and modifying its presented information according to selections made by the user. Users are interacting with the media to obtain the information they are interested in.

The common idea surrounding the idea of interactivity is that it is a quality of new media that allows the user some opportunity to change the nature or content of the information presented. In this sense the user is understood to be in some fashion 'empowered', elevated from the passivity of the existing media consumer. McQuail's (1986) assertion that new media afford registrational and conversational modes of use is supported by this view.

Manovich, however, critiques of the idea of interactivity. Manovich (2001: 55) initially contends that interactivity is a redundant term, as 'to call computer media "interactive" is meaningless – it simply means stating the most basic fact about computers'. He also contends that interactivity is a particular and strong discursive strand within internet studies. Manovich argues that this fascination is peculiar to theorists from within a Western tradition, as they regard the media as in some way representing the mental state of the user. A further critique is offered by Chapman and Chapman (2000: 14), who maintain that the idea of interactivity may be idealistic. They contend that human-computer interaction may not be true interactivity, as 'when the computer's role is to present choices and respond to them, it cannot be said to be keeping up its end of an interaction, while at the same time it reduces a user's options for contributing to the intercourse to a few mouse gestures'. Here there is some similarity with the description of interactivity defined by Downes and McMillan (2000). Human-computer interactivity is not regarded as full interactivity like that between two people. While a computer or forms of new media may be thought to be interactive, this definition must be differentiated from notions of communicational interaction.

Interpersonal communication

The second distinguishing feature of the internet also concerns the idea of interactivity. However, in this description interactivity refers to communication between people using the internet as a medium of communication. This is noted by Downes and McMillan (ibid.), and is recognised as a distinct theme in the literature surrounding interactivity. Included in this theme is a conception that users of new media may be widely distributed geographically, often across national borders. The distributed nature of users is understood to mean that audiences for certain types of content are organised in a different fashion to audiences for more 'traditional' media. Nowell-Smith (1991) has argued that traditional media have been understood as bounded by technological and political restrictions in terms of geography. Napoli (2003) argues that internet audiences are widely, if thinly, distributed. The idea of geographically distributed people communicating using the internet has received considerable attention in both academic literature and popular texts. The focus has been not only upon the understanding of the communicational potential of the media but also on the consequences of, or opportunities afforded by, its use. As noted previously, the ability of users to communicate with each other was a quality previously only attributed to one-to-one communication media such as the telephone. Being able to communicate with other users makes the internet a channel of both media consumption and communication. The interpretation of this quality in terms of the internet's impact on the formation of political identity has been significant, and will be examined in later chapters.

This quality of the internet, of affording interpersonal communication, is an area of incredibly rich popular and academic interest. Interest has been focused upon the various forms of communication permitted. For the most part these can be considered to be either asynchronous systems (communication that does not occur in real time) – e-mail and discussion lists, usenet and bulletin board systems (BBSs) – or synchronous systems (communication that is 'live' or occurs in real time) – text chat, multi-user domains or dungeons (MUDs) and virtual-reality graphical worlds. While initially much research focused upon text-based systems, more recently interest has shifted much more to the virtual graphical worlds, the growth in popularity of which has attracted interest from many different fields in academia as well as popular comment.

An early yet rich vein of research in this area centred upon the potential for subterfuge, deceit and game-playing that is understood to be possible when using such systems. The various chat and virtual-reality systems allowed a 'cues-reduced' method of communication to take place; the vast majority of 'normal' cues used in face-to-face communication – physical appearance, voice, gestures and body language – are redundant in these forms of computer-mediated communication. Instead, it is argued that the self-selected graphical representations or textual descriptions of users and lack of physical presence mean that users may not be quite what they present themselves to be.

Within academic research this possibility led to considerable research that sought to investigate the phenomenon of 'identity play' that takes place in these systems. While early work looked at deceit in interpersonal communications, using the internet in the presentation of an alternate gender to a user's 'normal' or 'real-world' gender, attention has also focused upon the relationships and social aggregations that emerge in such environments.

More recently the increased popularity of virtual worlds such as Second Life and the massive multiplayer online role-playing games (MMORPGs) such as Everquest has resulted in considerable greater popular and academic interest. While there is still interest in the earlier themes of identity, deceit, virtual communities and relationships, new issues such as the links between 'in-game' and 'real-world' economies (Castronova, 2006, 2007), the social interaction between players and emergent social dynamics (Kolo and Baur, 2004), among many other topics, have received attention.

Content production

The ability of users to produce content and for this content to be widely available to other users constitutes the third characteristic that differentiates the internet from older forms of media. The acceptance of this idea is shown very clearly in the huge growth in popularity of social networking sites, self-published media content sites and blogs. Social networking sites such as Hi5, MySpace, Facebook, Orkut and Bebo permit users to establish webpages, usually developed from a template, on themselves and invite other users to visit and link to their sites. The sites are a mix of various forms of asynchronous and synchronous communication channels and self-published content. While very much subject to fashion, social networking sites, part of the Web 2.0 portfolio of applications, have at least in the late 2000s (when this text was written) proven very popular. Moreover, this popularity is global in reach, at least in industrialised nations. Media content dissemination sites such as YouTube.com and MostPlays.com permit users to upload and easily disseminate a variety of forms of content, including video clips and self-produced games. The phenomenon of blogs must be considered at least as significant or more so than social networking and content dissemination sites. Blogs offer a means of special interest or even autobiographical publication. Users publish and frequently update content concerning their own field of interest and link to other blogs, news websites and other web content. While audience numbers for blogs are usually small, they provide a virtually free and accessible form of publishing requiring virtually no technical skill beyond that needed to use the internet. The significance of users being able to produce content is further demonstrated through the considerable industries of textbook and instruction book production, educational courses and technological provision in the field of web content production.

Underpinning these ideas is the perception that certain aspects of internet technology and software allow the production and publishing of content at little or negligible cost. Graham (1999: 69) contends:

> Compare the Internet with radio and television. Even in these days of infinitely many radio and television channels, and dramatically reduced costs of basic broadcasting, the possibility of individuals and small groups assembling the resources and know-how to put themselves or their views on air is still severely restricted, so restricted in fact that it is a practical impossibility for most. By contrast, individuals and groups with limited time, resources and skills can avail themselves of the technology of the Internet and, literally, present themselves and their message to the world.

Such an ability is a long-sought-after quality for more democratically orientated media theorists. The idea found early advocacy in the work of the playwright and literary theorist Brecht (1979: 169–71), who describes an idealistic use of the radio. Enzensberger (1970: 15) proposed a similar use of (then) new media and argued that new media would allow for the challenging of existing monopolistic systems of media.[2] Previous systems of media afforded only opportunities of reception; new media were understood to allow for production of media content outside of the media industry by individuals or groups: 'For the first time in history, the (new) media are making possible mass participation in a… productive process.'

Gauntlett (2000), along with numerous others, proposes that the internet in the form of the World Wide Web allows for such production. The web provides a system by which content can be produced and disseminated with considerable ease when compared to content production in older media forms such as television, film or print. The internet and, in particular, web technologies are understood to afford a system of media content production that lies outside mainstream industrial media content production. It is understood that users will be able to reach an audience using new media to which previously they would have had no access.

In addition to the commercial availability of software technology for the production of content, there has also been a conscious effort to develop operating systems and software for hosting, producing and consuming content that is assumed to be outside of monopolistic control, such as the Linux and Open Source projects.[3] This concept incorporates the idea that while user content may be produced outside the existing monopolistic systems of media content production, it is still subject to some corporate influence in the use of software and hosting systems. Accordingly, there is a strong and persistent line of argument proposing the use of non-corporate, subverted, free and shareware software. The production and dissemination of content by users with little financial and technological capital have been a foundation for the idea that the internet is an inherently democratising media form: a theme that will be explored in more detail in later chapters.

Individualised media

The fourth distinguishing feature of the internet is that it offers a more personalised form of media content. A chief advocate of this idea, Negroponte (1996), asserts that the digital nature of data and information passed by the internet means that content can be easily and cheaply copied, distributed and stored. The internet is regarded as being far more flexible than more established media in geographical and temporal terms. It is proposed that through systems of storage and retrieval the internet, in many instances, becomes a medium available upon demand rather than being available only at specific times (as broadcast media are understood to be). Furthermore, it is argued that new or digital media content can be disseminated across great distances and made available in degrees which old media, based upon physical technology, would find difficult to match. Negroponte (ibid.) describes the content of new media, such as the internet, as composed of 'bits' rather than 'atoms'. As the internet is available outside of the restrictive temporal and geographic specificity that binds old media, the amount of content available increases dramatically. The internet is understood to make available content that would be either temporally or geographically inaccessible under systems of old media distribution.

The greater temporal and geographic flexibility of the internet, along with the dramatic increase in the amount of content available and its unique human-computer interaction, mean that the internet may result in the much-touted personalised media (Gilder, 1994; Lasica, 2002a, 2002b) or 'Daily Me' (Negroponte, 1996; Spender, 1995), a long-established goal with new media advocates. The 'Daily Me' or 'personalised media' refers to collections of media content collated for specific users based upon user interest and preferences. Personalisation is defined as a 'recurring set of interactions between news provider and news consumer that permits you to tailor the news to your specific interest' (Lasica, 2002b). Personalised media are understood to allow users to define their own interests; accordingly, the media 'delivered' to them will be more closely targeted to their interests and lacking in redundant information and stories. According to Lasica (ibid.):

> Personalization is not just a cool feature of new media – it's intrinsic to new media. Unlike radio, television or print, the Internet is the only medium that is inherently personalizable. Users can be reached simultaneously with one-of-a-kind messages. The old formula of editors and news directors having the lone say in determining what's important has become an anachronism in cyberspace. The user, after all, is in the best position to know what he or she finds most interesting, valuable, useful – or newsworthy.

Lasica contends that flexibility in time and geography, and the resultant increase in content when coupled with user control, results in a technological system that challenges authoritative editorial decisions in the composition of media. To be able to circumvent the editorial power felt most strongly in old media is of considerable significance: it affords the internet a potency of anti-systemicism in addition to that usually attributed to the media. This is a topic that will be explored in more detail in later chapters. However, a less euphoric interpretation of this notion of 'individualisation' is also found within the literature. Commentators such as Stoll (1995) and academics such as Hargittai (2000, 2007) offer a more cynical and critical interpretation of the 'individualised' nature of the internet.

They have raised the issue that such systems, particularly web portals and search engines, may be operating with biases that influence the selection of content for users. These biases originate not in the user's preferences but in the initial programming of the software. Thus, while users may set their parameters for the retrieval of information, further influential systems of content selection operate. Furthermore, authors such as Jenkins (2006) argue that the individualisation of the media is not an end in itself but merely a phase, with personalised media being surpassed by a new understanding of media. In this new understanding, predicated upon the more 'user-active' systems of media in Web 2.0, media content is not individualised but rather produced and consumed (Jenkins argues the boundaries between production and consumption of media are blurring) by networks of users who benefit from the 'participation and collective intelligence' (ibid.: 245) the internet affords to users.

Conclusion

These ideas surrounding the internet indicate a general perception that such technologies are different from previously existing media forms. The consequences of understanding the internet as possessing additional qualities result, if McQuail's (1986) lead is followed, in new questions being developed concerning its use. Possible alternative theories and new approaches are thought to be required to appreciate its potential. To understand the nature of the approaches needed, McQuail's description of the internet, operating in the conversational and registrational senses as well as the allocutive and consultational senses, proves enlightening. Certainly, the internet can be considered still to operate extensively in the consultational and to a lesser degree in the allocutive senses. However, it has also begun to be understood as operating in the conversational and registrational senses.

The description of the internet examined here – the notion that it is distinct from existing media forms in four key areas: interactivity, interpersonal communication, user production of content and the individualisation of media – seemingly allows for it to be understood to operate in the last two of McQuail's categories of operation, conversational and registrational. The greatly increased potential for peer-level communication, the ability for users to communicate with one another and the possibility of centralised control of information used by individuals, as noted above, result in conceptions of use outside the more traditional understandings of media use.

The description of the internet as affording greater degrees of choice and decentred control over content, and the ability to disseminate content and communicate, all point away from an idea of the manipulatory power of the media and towards a conception of a media form that affords user empowerment and new forms of politics. The internet is widely understood to possess inherent qualities that allow for new forms of political activity. This perception proves of key import. It will be shown in later chapters how this view arises through interpreting the internet through a particular form of cultural understanding of the function of media. Furthermore, I will examine how the current interpretation of the internet needs to be understood not as universally applicable, but as grounded in political and social projects.

Notes

1 Memes 'are tunes, ideas, catch-phrases, clothes fashions, ways of making pots or of building arches. Just as genes propagate themselves in the gene pool by leaping from body to body via sperms or eggs, so memes propagate themselves in the meme pool by leaping from brain to brain via a process which, in the broad sense, can be called imitation. If a scientist hears, or reads about, a good idea, he passes it on to his colleagues and students. He mentions it in his articles and his lectures. If the idea catches on, it can be said to propagate itself, spreading from brain to brain' (Dawkins, 1989: 192).

2 Enzensberger's (1970: 13–14) categories of new media include 'news satellites, colour television, cable relay television, cassettes, videotape, videotape recorders, video phones, stereophony, laser techniques, electrostatic reproduction processes, electronic high speed printing, composing and learning machines, microfiches with electronic access, printing by radio, time sharing computers, data banks'.

3 The Open Source Movement (OSM), with its attendant agreements of the Open Source Initiative (OSI), presents itself as a collective of programmers who seek to produce software outside the proprietorial and restrictive system of commercial software production. The OSM operates a licensing system in which the source code of a program is available for inspection and adaptation. Use, including the packaging of future programs for commercial gain, of this source code can only be made if the resultant software is also OSM licensed. For a detailed analysis of the OSM see Wong and Sayo (2004), Raymond (1997).

4

The internet, politics and the public sphere

Introduction

Being the 'object' of study (rather than an academic subject with its own methodology and research agenda) means that the internet has been understood and examined by a variety of different disciplines. For some subjects it poses fresh challenges – there is currently a heated argument that media studies should be updated to deal with the challenges raised by new communications technology (Gauntlett, 2000, 2007; Merrin, 2008; Buckingham, 2008). For other disciplines the internet is regarded as an exciting new area of study: philosophy (Graham, 1999; Dreyfus, 2001; Cavalier, 2005), psychology (Gackenbach, 2006; Joinson, 2003; Wallace, 2001), philology (Crystal, 2006; St Amant, 2007) and even Isocratic rhetoric (Welch, 1999), to name just a few, have all sought to address the advent of the internet and new communications technology. In this chapter I wish to examine a further field of study – the way in which the internet has been understood in terms of its potential to affect politics. This is a large and popular field, and a strong argument has been made concerning the internet's power. Here I will examine just some of the ways in which it is thought to have impacted upon politics. However, I also want to make a further case: that just as the internet has affected politics, how we perform, carry out and understand politics has had a considerable effect upon how we understand the internet. In this and the following chapters I argue that the internet has been conceptualised as an inherently political medium, and that such an interpretation is built upon and incorporates a particular understanding of the media and political life in general – we have a view of the way in which politics works and how media interact with politics, and this understanding has influenced considerably how we view and understand the internet.

Examining the internet in relation to politics serves a further purpose – by integrating accounts of the internet and notions of political life we can begin to develop a socio-historical 'account' of the technology. This involves developing an appreciation of the nature of social life and the way in which such social life needs to be regarded as a key to understanding the possible ways in which the internet is used.

In examining the potential of the internet to affect politics it can be argued that three distinct strands or traditions of research exist. The first strand examines politics 'on' the internet; here the emphasis is directed towards actual activity occurring on the internet and its use for manifestly political reasons. Attention focuses upon how political parties make use of the internet for political communication and how certain beliefs are articulated in media content. The second strand is concerned with the effect of the internet as a medium upon politics. This examines the numerous ways in which the internet is thought to transform politics. The third strand addresses the politics 'of' the internet, and concerns itself with, for example, issues surrounding access, the political economies of production and the governance of the internet. This three-part distinction indicates a problem in much research upon aspects of political activity and the internet. The first and second traditions of research are underpinned by the central premise that technologies such as the internet are in some way 'outside' society and in a one-directional relationship with society. Moreover, the focusing of research on politics conducted on the internet, that which occurs online, and how the internet will affect politics and political activity is further at fault in that it delineates activity conducted 'online' from wider issues, such as access to computer and internet facilities and the possession of the intellectual capital that may directly affect online activity. As Slevin (2000: 6) notes, the internet constitutes a 'modality of cultural transmission', one of the many ways or 'locations' in which communication may occur. Communication that takes place online should not be divorced from activity that takes place away from the internet. Online communication constitutes just one of the many forms through and by which mediated communication takes place. This chapter, therefore, will examine the ways in which the perceived qualities and the general understanding of the internet have been seen to result in new forms of association and communication. More particularly, this chapter will be concerned with how these new forms of association, seemingly enabled by the internet, have been understood in terms of their potential to change or bring about political action.

I argued in the preceding chapter that the internet as a medium is thought to possess a set of unique qualities that distinguish it from other media. These qualities or characteristics allow the internet to be used in ways in which old media simply could not. The potential afforded to the individual by the internet to develop a more individualistic or personal pattern of media consumption and the possibility to produce and disseminate media content have been understood to induce a number of changes in the relationship of media and politics. Of particular note is the contention that the internet offers the opportunity to re-establish the public sphere and enable new forms of politics. Such 'new spaces' brought about by the internet are understood to be imbued with a radical potential, and consequently the internet is often regarded as a medium that can be used in an 'anti-systemic' manner; it is a counter- or small medium that may provide a voice to those disenfranchised by the mass-media forms. Furthermore, the internet is believed to offer the opportunity for the establishment, expression and deployment of alternative or counter-hegemonic forms of identity. Such dual anti-systemic action has been understood in a variety of ways, ranging from a corrective force acting for the re-establishment of democratic forums to a challenge to the pathologic incursions of state or commerce into public discourse, or even to a truly radical medium, affording revolutionary potential. Here I want to focus attention upon one key way in which the internet has been seen to bring about change of a political nature: its ability to revitalise the public sphere. In exploring this issue I look at how the public sphere has been conceptualised and how it is regarded as being 'revitalised' by the communication forums possible on the internet. Indeed, the internet often seems to be regarded as a saviour of the public sphere.

New forms of communication and the public sphere

A key facet in the argument surrounding the idea of the internet and political activity is that the widespread deployment of the internet will result in a shift in power away from the institutional producer and towards the consumer or user. McQuail (1986) notes how the use of media that function in a non-allocutory or non-consultory sense may result in differing patterns in the operation of power: media that afford the user control over the flow of information at point of use and the opportunity actually to produce publicly consumable media content may result in a system that operates in a centrifugal fashion, dislodging power from a central position to outlying regions of civic life. Such arguments have been strongly present virtually since the inception of the internet, or at least since it achieved a degree of popularity in the 1990s. The various systems of interpersonal communication used on the internet – e-mail, newsgroups, the many chat systems, the various forms of virtual community, virtual reality, massive online games and more recently blogs and social networking websites – have been interpreted as having an inherently radical potential, offering new spaces and forms of communication that may allow for the restoration of the public sphere of democratic politics.

The idea of the public sphere was popularised by (and in response to) the work of the German philosopher Jürgen Habermas. Briefly put, the public sphere is an idealised, virtual or imagined space in which members of a community may communicate. The public sphere does not necessarily exist in a particular geographical space; rather, it is a concept or component of a particular political system. Habermas regarded the public sphere as a concept in the practice of democracy that was singularly identifiable in a particular historical, political and social situation. It is certainly not a constant feature of all human social life, and was deeply linked to certain social conditions that afforded its emergence and eventually caused its demise. Habermas developed his ideas in several major works, commencing with the publication of his 1962 *habilitationsschrift* (post-doctoral research thesis produced to gain an academic position) titled *Strukturwandel der Öffentlichkeit. Untersuchungen zu einer Kategorie der bürgerlichen Gesellschaft*. This was subsequently published in 1989 in English as *The Structural Transformation of the Public Sphere: An Inquiry into a Category of Bourgeois Society*. It was further developed in a number of texts (Habermas, 1985, 1989, 1996, 2004, 2005, 2006). Habermas offers an account the emergence (during the late eighteenth century) and later demise of the public sphere, the 'sphere between civil society and the state, in which critical public discussion of matters of general interest was institutionally guaranteed' (McCarthy, 1994: xi). Habermas proposes that in early modern life there came into existence new forms of 'public life'. This 'public sphere' was a reaction to the monopolistic and absolutist control of political life by royal courts, and involves the convergence of an emerging bourgeoisie and a section of the aristocracy separating from the royal court. Such new forms of civil association were, according to Habermas, to be considered the articulated rational opinions of élite private citizens. The opinions of such private citizens, when expressed through publicly available media such as printed booklets or in public spaces, allowed for the formation of a body of 'public opinion'. This public opinion replaced the existing situation, in which the power of the ruler was simply imposed upon the populace.

Habermas's concept of the public sphere is a multi-part account of political life, and he divides the 'social world' up into distinct areas. The first area is that aspect of life in which we deal with 'officialdom' and the systems that make life in advanced capitalist societies possible: the laws and regulations, the various systems that we must live by. This aspect of life is governed by a particular form of administrative or economic logic or 'rationality'. Outside this economic and administrative world there exists a 'life world' based upon a different form of rationality. Here the rationality or logic used to judge actions is more transparent and open, and decisions are based upon a different set of ideals or 'ethics'. It is in the 'life world' that people for the most part live, and within it are what are termed the 'public' and 'private' spheres – the spaces of individual and social action. This notion of 'civil society', a realm beyond the merely economic, emphasises shared political effort and social organisation outside the traditional political system. The public and private spheres are conceived of as separate from the economic world.

This model is based upon an account of European social and political history, and Habermas sees the advent of universal suffrage and mass political participation resulting in a gradual transformation of the public sphere. Indeed, it was the culmination of the very process that enabled the public sphere that eventually destroyed it. Problems related to the new hierarchies based upon wealth, the extension of the state into numerous areas of public life and the rise of the power of the mass media resulted in a crisis for the public sphere and a gradual 'refeudalisation' of social life. Developing further the Marxist critique of the mass media prevalent within the Frankfurt School, the growth of mass communication is understood to contribute significantly to the decline of the bourgeois public sphere. Whereas the embryonic mass media had initially functioned as a facilitator for the emergence of the public sphere, the media's later corralling by conglomerates and state interests resulted in their becoming agents of control, and systemically formulating public opinion for the benefit of an élite minority. Habermas proposes that the solution to this dilemma of the decline of the public sphere lies in the salvaging of rational discourse and the general cleansing of society-wide communication pathways. His work upon the development of what he termed 'discourse ethics' – the sets of standards by which decision-making processes can be regarded as legitimate by those whom they affect – helps to identify and examine possible instances of the public sphere.

Habermas's work has been subject to substantial criticism and development (Calhoun, 1992; Fraser, 1992a, 1992b; Benhabib, 1996; Mouffe, 1999). The idea of a sphere of activity separate from other elements of life proves difficult to substantiate. Fraser (1992a, 1992b) notes the necessity of the exclusive nature of the public sphere. Only through exclusion of the proletariat and of women and in its distinction from popular culture could the public sphere maintain its claims of rationality – the sphere was never truly public, and facilitated communication only for a new section of society, the bourgeoisie. The 'logo-centricity' of the public sphere, its relegation of other or 'non-rational' forms of discourse, ensured its exclusive and élitist membership and proves difficult to defend. Furthermore, as Mouffe (1999: 757) points out, the idea of a singular public sphere is problematic. A public sphere must accommodate the 'multiplicity of voices that a pluralist society encompasses'. Accordingly, the possibility of ideal speech occurring in a bourgeois public sphere is challenged. Ideal speech, meaning equal participation in a debate between individuals, is impossible in a pluralist society, and communication will always be between unequal discussants.

Habermas's later, post-1989 work sought to remedy the inadequacies of his initial model by acknowledging the possibility of a multiplicity of spheres rather than a singular space and establishing discourse ethics as guidelines. Habermas's model is firmly progressive, and seeks to salvage certain core aspects of the project of modernity that were seen to have been so badly savaged by twentieth-century history. The debate surrounding the publication of his work has resulted in a renaissance of work concerning the notion of the public sphere, public space, the role of the state in communication and the nature of political discourse.

The internet and the public sphere

Some of the most marked modifications and adjustments to Habermas's theories have arisen in response to changes in advanced capitalist societies since 1989. As mentioned above, Habermas acknowledged certain problems with his initial description of the public sphere and made a number of modifications to his theory. In addition to the recognition of a plurality of public spheres, a number of authors have sought to utilise the notion of the public sphere in examining the internet.

One early yet still strong strand of thought regards the internet in a primarily beneficent light. This strand of thought, commonly termed 'cyber-utopian' (Turner, 2006), integrates a strong belief in the power of technology with progressive counter-cultural thinking. Regarded by Carey (1989, 2005) as espousing the rhetoric of the 'technological sublime', cyber-utopians see the internet and associated communicative systems as having a very positive effect and assisting in the protection (Arterton, 1987) or restoration (Barber, 1984) of democracy. Often articulated by activists and others concerned with progressive issues, cyber-utopians regard the internet as a means by which social projects may be achieved. For example, Walch (1999) argues that the internet may help to strengthen democracy and invigorate civil society. Rheingold offered a similar yet more measured work on the virtual community, initially published in 1993 and reissued in 2000. In this text, as in his public speeches and blog postings, Rheingold is concerned with whether the internet can contribute to the public sphere. Articulating a strongly positive answer to this, Rheingold argues that the internet and the virtual communities enabled by it could lead to a rebirth of populist democracy and have an immense transformative power in assisting in the restoration of the public sphere. Indeed, Rheingold (2008) is critical of more reserved interpretations of the internet, and goes so far as to argue that Habermas 'doesn't understand the phenomenon he is describing'. It could be more justifiably argued that Rheingold does not entirely understand Habermas.

This question of whether the internet facilitates the public sphere has attracted a lot of attention from researchers. Dahlberg, who is cautiously optimistic about the power of the internet to enhance the public sphere (Dahlberg, 1998, 2001a, 2001b), identifies a distinct trend in research that he refers to as 'deliberative democracy' (Dahlberg, 2001b, 2001c, 2007). This position proposes that the new technologies offer a means of rectifying the distortions to forms of communications from intrusive state bodies and commercial interests, and a reassertion of the bourgeois public sphere and idealised speech communities of early modernity. Older forms of media have seemingly failed to deliver the public sphere (or have been so compromised by state or commercial interests as to be harmful to the public sphere). This argument is developed by a number of authors who see the internet as able to facilitate communication between individuals and encourage discussion in a way that older existing media could not. Stromer-Galley (2003) details a strong belief among participants that the internet, and in particular newsgroups, facilitates a rich diversity of opinions. Boeder (2005) regards speculation on the demise of the public sphere as premature, seeing the internet as a possible source of strength. Similarly, Gimmler (2001) regards the internet as well able to contribute to the public sphere and even to direct political debate. Froomkin (2003) identifies one particular example of communication that took place on the internet: the discussion regarding the development of technical standards. Froomkin argues that this process of developing standards may be considered an exemplar of discourse ethics in action, though he acknowledges that it is but a minute aspect of internet communication and the rest of the internet may well not meet the requirements of or be considered a public sphere.

Poster (1995, 1996) offers a critical examination and asserts that the internet may afford an opportunity to salvage aspects of the bourgeois public sphere. Additionally, the decentralised nature of internet communication confers upon users a chance to engage in communication outside the dominant systems of media hegemony. However, interpreting critiques of Habermas's description of the public sphere, Poster retains a degree of critical awareness in promoting the internet. The post-modern subject, Poster asserts, is not the liberal subject of Habermas's original public sphere but an assemblage of multiple discursive roles. A number of other authors have developed a similar notion, though without the adherence to post-modern theorisation. Dertouzos (1991) asserts the internet will ensure that users of such technologies will be able to express opinions freely. Similarly, Kellner (1998) notes the operation of the internet as a place of effective oppositional action within the American intellectual community. The internet offers marginalised intellectuals, those denied access to more mainstream forms of media production, a means by which they may engage in debate. A strong example of the 'deliberative democratic' position is offered by Bryan et al. (1998: 5), who précis the arguments of the 'civic networking movement' in asserting that:

> New media, and particularly computer-mediated communication, it is hoped, will undo the damage done to politics by the old media... new media technology hails a rebirth of democratic life. It is envisaged that new public spheres will open up and that technologies will permit social actors to find or forge common political interests.

In another text in the same volume Tsagarousianou (1998) mitigates such euphoric and determinist claims by asserting that local political and cultural considerations play a key role in the process of technology enabling political action.

Particular communicational technologies within the internet have also been identified as being strongly tied to the redevelopment of the public sphere. Much early work focused upon the use of newsgroups in relation to political discussion. More recently, blogs have attracted a great deal of attention. In a similar way to how newsgroups and other preceding forms of computer-mediated communication were understood, blogs are credited with an enormous transformational potential. Central to such claims are three key arguments. First, blogs encourage civic participation and involvement – people can be re-empowered and civic culture revitalised (Blood, 2002; Colvile, 2008). Second, blogs revitalise politics and bring in a new level of accountability for politicians (Kline, 2005; Colvile, 2008). Third, blogs erode the influence and power of formal groups and existing power structures (Hewitt, 2006) and challenge corporate media power (Gilmor, 2006; Scott-Hall, 2006).

In a less euphoric tone Barlow (2007) argues that blogs do not present a new form of journalism but represent (and restore) a theme of citizen journalism in American political life. This idea that blogs continue a liberal and distinctly American form of communication and political action is important. As will be explored in later chapters, it is the interpretation of the internet through a lens of primarily American, and to a lesser extent Western, understanding of the role of media in political life that influences how we understand the internet. Continuing this more critical account, a number of authors have theorised blogs within a model of the public sphere (Froomkin, 2003; Ó Baoill, 2005), though for authors such as Keren (2006), celebrating blogs for their potential to rejuvenate the public sphere may be a little premature.

Other authors have shied away from such direct and progressive linkages between the internet and a resurgent public sphere. Wilhelm (2000: 6) seemingly reasserts the substantivist arguments when he asserts 'new information and communication technologies, as currently designed and used, pose formidable obstacles to achieving a more just and human social order'. A further criticism is the identification of a number of limitations imposed by commercial interests upon online communication that restrict and challenge open debate. Phenomena such as overt commercial interest stifling free comment (Dahlberg, 2005), commercial influence affecting information retrieval (Hargittai, 2000, 2007), overt regulation of internet content consumption (Gomez, 2004) and monopolistic ownership of media organisations (Bagdikian, 2004) all pose challenges to the idea of unfettered communication. Further, issues of access and skills deficits (Selwyn, 2004) and power relations within online groups (Dahlberg, 2007) also challenge the link between the internet and public sphere.

Perhaps the greatest barrier to considering the internet as an actualised public sphere lies in the fragmentation and partisanship of online communication. Internet 'audiences' are notoriously partisan; indeed, as Dalhlberg (2007) notes, certain aspects of the technology seem almost to encourage users to stick to particular sources. Features of software such as 'favourites' or 'bookmarks' make returning to the same webpage very easy in browsers. Moreover, research indicates that even in deliberative media such as discussion groups, blog links and mailing lists a high degree of insulation from opposing views occurs. Participants in online deliberative environments tend to stick within their community of interest (Hill and Hughes, 1998; Wilhelm, 1999), 'pockets of interest' (Selnow, 1998) or 'deliberative enclaves' (Sunstein, 2001). The mutually supporting belief groups may seem to exacerbate differences and lead to extremism, a far cry from the development of citizenship and rational discourse that Habermas (1996) identified as the goal of deliberative action. Indeed, Habermas places little faith in online environments to facilitate the public sphere. Instead he sees them as dangerously fragmenting of politics and at best as little more use than ancillaries to mass media in the development of civic discourse:

> the rise of millions of fragmented chat rooms across the world tend instead to lead to the fragmentation of large but politically focused mass audiences into a huge number of isolated issue publics... the online debates of web users only promote political communication, when news groups crystallize around the focal points of the quality press, for example, national newspapers and political magazines. (Habermas, 2006: 423–4)

While the internet may be considered as detrimental to a traditional or nationally orientated public sphere, it may possibly afford opportunities for the development of other forms of public sphere. Keane (2000) offers a multidimensional model drawing upon many of the criticisms mounted against Habermas's original account. Keane proposes that among other changes the development and widespread deployment of the internet within Western societies, and less so globally, have resulted in the emergence of three forms of public sphere operating at different distances from the individual.

Keane uses the term 'macro-level' to describe communications technologies such as certain internet sites and satellite broadcasting and media structures of the pan-national media organisations. For example, through their websites the BBC and Fox News permit a similar media experience to be shared by tens or hundreds of millions of individuals across a wide range of societies and cultures. This contention is also supported by the notion that internet audiences may be widely yet thinly distributed. Such macro-publics afford the possible integration into a single 'public of choice' of geographically disparate individuals. Keane (ibid.: 83) contends that 'There is a category of users with a "net presence" who utilize the medium… as citizens who generate controversies with other members of a far flung "imagined community" about matters of power and principle.' Keane cites the Association of Progressive Communications as one example of such a geographically disparate community.

The second form of public spheres Keane refers to, 'meso-level' spheres, are a less recent historical phenomenon and consist of publics typically bounded by political or state boundaries. They also include those publics unified by a single self-defined notion of culture or ethnicity but distributed across often bordering states. They may further be notional minority nations within states. Typically these publics are served by more traditional forms of media, such as newspapers. Such forms of public, enabled by systems of mass media, have a significant history and are of considerable longevity.

The third form in Keane's model, 'micro-level' public spheres, are understood as those groups operating along new forms of social allegiance. Melucci (1980) has described the (relatively) recent historical genesis of 'new social movements', such as the women's movement and the green movement. Melucci has argued that such movements indicate a growing retreat from traditional forms of identification and the emergence of counter-public spheres within existing more micro-level spheres. Keane argues that the internet, among other technologies, enables such micro-spheres.

The descriptions of the internet noted above, be it either enabling or fracturing the public sphere, seem to fit well with Keane's model. The internet affords action across national boundaries, but also among small interest-oriented groupings with alternative methods of identification to that found in the meso-sphere. As Villarreal Ford and Gil (2000: 203) contend, the internet permits such spheres as 'it consists of people's participation in creating interactive forms of communication that act as a countervailing force to the one way flows inherent in commercial media'. The interactive and interpersonal potential of the internet offers considerable scope for the articulation of politics that lies outside of those forms that receive the majority of attention in mainstream media, typically offered by the meso-level Keane identifies.

In spite of the reservations concerning the corporate influences upon it and its fragmenting nature, the internet is viewed as having a significant potential for the re-establishment of the public sphere on both micro- and macro-levels. Esteva and Prakash (1998) note that the formation of such new spaces is dependent upon the participation of entities separate from formal state institutions. Such civic organisations will be composed of 'autonomous, democratic civil society as it expresses itself in organisation independent of the state and its formal corporate structures' (ibid.: 11). The internet is understood to afford an opportunity for new space, a space separate from formal, state-sanctioned areas.

The counter-public of the internet?

In spite of the reservations noted previously, for the most part the new forums and means of communication brought into being through the internet are not regarded as coercive or compromised. Indeed, a number of qualities of the internet are assumed to allow for a counter-position to the dominant form of belief to be articulated. Downey and Fenton (2003: 196) describe such a position as a 'counter public' and contend that 'recently... the internet has been hailed as the saviour of alternative or radical media and indeed politics, perfectly matched for the widely-dispersed resistance of culture jammers and radical political protesters by both theorists and activists'.

The internet, or at least certain communicative aspects of it, is understood to be a media form outside the dominant system of media ownership and control (a position strongly challenged by authors such Bagdikian, 2004). The internet, with its alleged inherent ability to afford individuals the ability to communicate with one another, is believed to offer opportunities for, and even instigate, oppositional activity.

81

Such is the potency of the internet as a possible counter-public that its use segments users from the normal discursive realm of hegemonic media consumption and situates them in a new and radical environment. Villarreal Ford and Gil (2000: 223) argue that:

> People who participate in posting and debating on the internet occupy a discursive realm outside of mainstream media. They may speak freely and still enjoy a wide audience, a remarkable opportunity in a world in which information and its means of distribution are so closely guarded by politicians and corporate interests.

A number of authors continue this progressive reading of the internet in relation to counter-public spheres. Sassi (2001), for example, describes the potential of emerging political entities, operating solely through new media technologies, to offer an alternative to troubled 'real-world' political entities. This conception of the media is not recent in origin. Enzensberger (1970), in a comparatively early and prescient paper, described how new media had the potential to 'mobilise', to disseminate anti-systemic information to people via systems that were outside the corporate or state-owned media. Sreberny-Mohammadi and Mohammadi (1994) talk of the revolutionary potential of 'small media' and the ways in which non-hegemonistic media offer the opportunity to articulate discourse that is outside mainstream media. They investigate the potential of small media to mobilise and motivate people towards revolutionary action. Walch (1999) is much more explicit in examining the potency of the internet. He argues that various groups have, under the auspices of the Association of Progressive Communications, been established to use the internet and computer networking technology to communicate outside the mainstream forms of communication. The internet is seen to be offering alternative channels of (political) communication that can be used to challenge the orthodoxy of hegemonic discourse. The spread of computers and the establishment of the internet 'opened a door for non-hegemonistic individuals and groups to counter the consciousness-creating hegemony of the cultural and informational elites' (ibid.: 49).

Conclusion

The intention of this chapter has been to sketch some of the ways in which the internet has been understood to intersect with politics, and in particular the public sphere. The internet's potential to actualise the public sphere has been much heralded, but so have the numerous barriers that may stop it functioning in that manner. While the internet certainly has offered new opportunities for the expression of opinion, the problems that beset earlier forms of communication are shown to be just as problematic on the internet. The power of corporations and state bodies is still felt in the various communicational channels of the internet. Moreover, existing and new political hierarchies are manifest. Simply because 'presence' is lacking does not prevent power from being expressed in online environments. Perhaps most problematic, however, is the insulated and fragmented nature of the communication that does take place. As noted above, research reports that many users stick to what they know, rearticulating already-held beliefs. In this way discussion reinforces rather than challenges beliefs. If such discussion does not assist in the facilitation of the traditional form of the public sphere, it is thought to assist in developing counter-spheres or places where contrary opinion can be voiced. Here the internet's quality of affording communication between thinly distributed but like-minded individuals is seen as a distinct strength. From such a perspective the internet is viewed as imbued with a democratic political potential, and while such a quality has also been detected in other media forms, such as Samizdat[1] and other 'underground' publications, this democratic quality is regarded as intrinsic to the internet. The internet is a media form imagined with an inherent and essential democratic potential.

However, we must raise a note of caution here. Such a 'reading' of the internet is possible only because the internet has been conceptualised within a particular model of media and politics. We have a 'template' concerning the relationship of media to politics, a set of arguments that have already been established. We simply 'slot' the particular media form under discussion, in this case the internet, into this template and are able to understand it, or at the very least debate its usefulness in bringing into being the public sphere. Put crudely, we see the internet as contributory (or not) to the public sphere because our model of politics says that is what a medium will do. We understand media will operate in a particular way, and apply this model to any and all media. It is the particular mode or interpretation of the media and politics and how the internet has been conceptualised within it that will concern me in the next chapter.

Notes

1 Samizdat (Russian for 'self-publish') is a historical publishing form dating from the USSR in the late 1950s. It initially

referred to literature that, due to censorship restrictions, could not be published through official channels. Telesin (1973: 25) notes how an author in 1950s' Russia would have, in fear of reprisals for seditious poetry, 'without waiting to be published (or, perhaps, in despair), bound together the typewritten sheets of his poems and wrote *samsebyaizdat* where the name of the publishing house normally appears in a book', *samsebyaizdat* being an ironic play upon the name of official publishing sources of Politizdat (political publications) and Yurizdat (legal publications). Samsebyaizdat was shortened to Samizdat, and became associated with publications that circulated via being either hand copied (referred to as 'overcoming Gutenberg') or typed using carbon paper (referred to as being published by 'Underwood', after the typewriter manufacturer). It eventually came to refer to any illicit publication or magazine within totalitarian regimes.

5

Politics and the role of the internet

Introduction

This chapter examines how the dominant, accepted and for the most part largely uncontested description of the internet, detailed in the preceding chapters, has been understood through, and influenced by, systems of political thought. The usual description of the internet, a medium that is firstly clearly distinct from other forms of media and secondly a politically powerful medium, emerged within certain theoretical political frameworks and worldviews. These frameworks incorporate specific perceptions of the social and political world, of the nature of 'public space' and of the media. Accepting these notions of democracy, the public sphere and the individual underpins or provides a basis from which we understand, interpret and examine the internet. Moreover, these frameworks have become normative in accounts of the use of the internet: they influence the way in which the internet is described and how its power is discussed. In brief, how we understand politics plays a considerable part in how we have gone about examining, interpreting and thinking about the internet and its intersection with society.

In the concluding remarks of the preceding chapter I argued that the internet has often been understood as being used in a radical or contrary sense to other media, and that such a conception depends upon the acceptance of certain political assertions. It is these political assumptions that are discussed in this chapter. In exploring this issue I will examine two contrasting perspectives found within discussions of the internet. The first perspective I will describe as 'liberal democratic'. This is fundamentally a 'category name' for a range of pluralist political theories centred on ideas of democracy. Political and academic critiques of positions within this perspective have been wide-ranging and are collated here under the title 'critique of existing systems'. Arising from this critical reading of liberal democratic theories, a further perspective becomes apparent, referred to here as 'radical democratic'. This perspective is often used within academic and critical discussions of the mass media and the internet.

Conceptualising perspectives on the media

In the preceding chapter it was noted how Habermas's concept of the public sphere has been challenged. These challenges, and the various positions taken on the media and the role of the public sphere by different theoretical perspectives, were examined by Curran (1991). He offers a description of the key perspectives and concepts involved (Table 5.1).

Curran's table describes the ways in which various aspects of the media and the public sphere have been interpreted by different schools of thought. Understandably, due to its age,

Table 5.1 Curran's model of perspectives on the public sphere and the media

	Liberal	Marxist critique	Communist	Radical democratic
Public sphere	Public space	Class domination	—	Public arena of contest
Political role of media	Check on government	Agency of class control	Further societal objectives	Representation/ counterpoise
Media system	Free market	Capitalist	Public ownership	Controlled market
Journalistic norm	Disinterested	Subaltern	Didactic	Adversarial
Entertainment	Distraction/ gratification	Opiate	Enlightenment	Society communing with itself
Reform	Self-regulation	Unreformable	Liberalisation	Public intervention

Source: Curran (1991: 28)

there are a number of significant problems with this description. Notably lacking are critical theories other than Marxism – the concerns of feminists, gender theorists and anti-racist or post-colonial perspectives are all absent. Furthermore, there seems only a partial acknowledgement of the developments in social theory of the media reflecting the introduction of post-structural theories. Table 5.2 is a revised version of Curran's model.

This table differs in a number of respects. First, the topics covered are revised and reduced in number to four. The first is the role of the media and concerns the function of the media in differing models of society. The second topic is the public sphere, and refers to both the collected opinions of a notional society and ideas of opportunities or possibilities of political action, commonly understood as 'political space'. The third topic is the control of the media, and concerns the

Table 5.2 Revised model of perspectives on the public sphere and the media

	Liberal democratic	Critique of existing systems	Radical democratic
Role of media	Fourth estate, ensuring behaviour of state	Opiate, propaganda, legitimating and oppressive through exclusionary and constructive practices	Often hegemonic and coercive
Public sphere	Public space, site for legitimate expression of opinion	Bourgeois, heterosexual, male, white, discourse predominantly exclusionary of counter-discourse	Area of contest
Control of media	Market/self-regulation	Bourgeois, male, heterosexual, white-dominated	Public service
Conception of internet	Small media		Radical media

best means to regulate the ownership, content and control of the media. The fourth topic is the differing ways in which the internet is conceptualised and interpreted. Second, the perspectives on the media are revised. The liberal position is retained and expanded to one termed 'liberal democratic'. The critical position in Curran's model is extended to include the feminist, gender theorist and post-colonial critical positions under the title of 'critique of existing systems' – I do acknowledge, however, that grouping together Marxist, feminist, gender theory and post-colonial positions is a significant reduction and perhaps a gross misarticulation of their perspectives and interests. As with Curran's model, the table also includes a description of the 'radical democratic' position, a stance that derives from a critical reading of the media. Included in the critique of existing systems' position and the radical democratic position are concerns arising from the development of post-structural theories, partially inseparable from developments in feminist, gender, post-Marxist and post-colonial theory.

The liberal democratic perspective

The perspective referred to as liberal democratic is a broad range of theories that can be found not only in pluralist and journalistic accounts of the role of the media in public life but also within more conservative accounts. Although there appears to be considerable difference between the various schools of thought in such a perspective, they actually share many common ideals. This is particularly evident when they are contrasted with more 'radical' perspectives. While disagreements over the mechanisms of the means of distribution (markets versus regulation) distinguish the different political stances in most Western democracies, such debates mask a basic acceptance of the liberal democratic doctrine of freedom, equality, legal systems of conflict resolution and acceptance of legal safeguards of private property. Truly contrary opinions are rarely, if ever, voiced in mainstream political debate, where the vast majority of opinions lie within the remit of liberal democratic practice.

Key to the liberal democratic position is an assertion that mass media fulfil a specific role in the operation of democracy. Curran (2000: 121) notes that a key role in the liberal understanding of the media is their ability to 'act as a check upon the state'. The media are a 'fourth estate',[1] a component of the political system outside the official administrative realm. Privately owned or independent mass media are understood to function as a guardian of citizen interest in the operation of political power, monitoring, reporting on and chastising the activities of the political élite in democracies. The potency of the press to affect the political process, to safeguard the interests of citizens in the face of malevolent political power, situates the media as a vital part of the democratic process. Carlyle (1859: 164) argued that this power was derived from printing, which in turn emerged from writing, and claimed a direct link between writing and democracy: 'Invent writing, democracy is inevitable.' This is a point developed by Keane (1991: 7), who notes that such a conception of the media has a long history within Western Europe and the USA. Keane also notes how the historical development of such an ethos is peculiar to these regions, a point that will be examined in later chapters.

It is worth noting that older accounts such as Carlyle's often consciously disregard ideas of social class; these aspects are not considered important in the individual's potential to speak to the broader constituency. Carlyle (1859: 147) contends:

> Whoever can speak, speaking now to the whole nation, becomes a power, a branch of government, with inalienable weight in law-making, in all acts of authority. It matters not what rank he has, what revenues or garnitures: the requisite thing is that he have a tongue which others will listen to; this and nothing more is requisite.

Social class is not conceptualised as a barrier to expressing one's opinion in public. However, the ability to speak in public, to engage in public debate, seems contrary to the nature of the public sphere described by Habermas. For Habermas, as noted in the preceding chapter, the historic public sphere was composed of élite members of society; but the liberal democratic conception regards the public sphere as the collective opinions of all members of a society, expressible through mass media. The liberal democratic public sphere is the totality of communicative space existing in the 'public' or social world. This implicitly incorporates a notion that some aspects of the social world are not public but private, a point accepted by Habermas's rationale. Public space in the liberal conception is communication intended for a non-local audience, communication beyond the 'personal'. Fraser (1992a), and later Habermas (1992), noted that the public sphere described in Habermas's early work was composed of members of an élite social grouping, and as such there were numerous individuals outside this who were unable to participate. The public sphere is the sphere of action that, while notionally open to all, is in essence exclusionary of those who exist entirely in the private sphere; it is the world outside of the private, a 'second space' of action. Furthermore, the public sphere for liberal democratic theorists provides an arena in which political action may legitimately take place. Public life is where 'politics happens', it is the space in which contested issues are legally resolved and debates settled. There is a direct equation between politically legitimate action and that action which occurs in the public sphere. It is worth noting, as Cammaerts (2007) does, that US and European concepts of deliberative practice and public life may differ slightly.

Debates within liberal democratic political positions concerning the regulation of mass-communications media have tended to oscillate between notions of market freedom, public state regulation and independent or non-state regulation. The current ascendance of various flavours of market and self-regulation is by no means a permanent feature of liberal democratic political discourse, and may well be waning. Additionally, authors such as Steemers

(1998), Tambini et al. (2007) and Terzis (2008) have noted that the introduction of various forms of new media may also bring about forms of change in regulatory practices. Implicit in all contemporary accounts is a historic tension between the conception of the freedom of the media (inclusive of their role as a guardian of the interest of citizens) as an irreducible right, as noted above, and the necessary regulation of the media for societal benefit. It is movement along this continuum that affords the articulation of contrary opinions and political positions within a broad framework.

The internet as small media

As noted above, the public sphere forms the arena in which legitimate political activity takes place. Accordingly, political action that occurs outside the public space is not deemed legitimate and is conceptualised in a variety of ways, such as subversive, traitorous, non-political or even criminal. The use of various media forms to engage in such illegitimate political action is viewed as inappropriate. Typically, the use of the media to disseminate radical ideas is subject to legal sanction – the activity being designated as criminal, or at least a civil offence. Within liberal democratic systems of thought there are two areas of concern. Firstly, there is contention as to whether certain content is acceptable as publicly disseminable within the public sphere, or whether such content constitutes extra-public or non-sanctioned discourse. Secondly, the nature of the media used to convey meaning is also contentious. Certain forms of communicative apparatus are considered legitimate, such as radio and television broadcasts by state-sanctioned operators, for example, while others are illegitimate and non-legal, such as graffiti and pirate radio (Downing, 2001).

The stringent defence of various communicative practices occurring on the internet lies, however, not in the claim of presumed legality but in the resistance to the possible incursions into already legally accepted communicative practices. The internet is understood as a legitimate form of communication and, barring the occasional occurrence of illegal content, content is by default legal in Western democracies. Contention occurs when attempts are made to intercede and monitor communication occurring on the internet. It is the defence of the right to communicate unhindered by the attempts of the state (imagined or real) to intervene that concerns activists in this field (Barlow, 1996; Cammaerts, 2007).

Accordingly, the internet cannot fully be regarded so much as a truly radical media form, though it is often regarded as being highly unconventional. From this perspective certain forms of internet communication should instead be regarded as a form of 'small' media. This is a term derived from Sreberny-Mohammadi and Mohammadi's (1994) description of certain media. Small media are the forms that exist outside major commercial media industries but still within the totality of legal systems. They are defined as 'participatory, public phenomena, controlled neither by big states nor big corporations' (ibid.: 221). Small media bring attention to:

> an autonomous sphere of activity independent of the state, the popular production of messages, a public coming into being and voicing its own 'opinion' in opposition to state-orchestrated voices; to the use of channels and technologies that are readily accessible and available; and to messages that are in the main

produced and distributed freely as opposed to private/corporate production for profit or control by state organisations. (Ibid.)

Small media afford the articulation of contrary discourse, but as Moore (1978) asserts, 'for any social and moral transformation to get under way there appears to be one prerequisite that underlies all... *social and cultural space* within the prevailing order' (emphasis in original; cited in Sreberny-Mohammadi and Mohammadi, 1994: 223).

Small media, such as the internet, offer an alternate channel for the articulation of discourse, a channel that while legal and legitimate offers an space for non-institutional and non-corporate voices. The characterisation of small media proposed here means that the internet is not truly a radical media, but is instead a means by which individuals who were excluded from the public sphere and public life are provided with a method to re-enter it and add their voice to public life.

Critique of existing systems

As noted above, the critique of existing systems incorporates a range of critical positions on the media; they are, perhaps somewhat unjustly, merged together into a single category here. The breadth and depth of critical analysis of the media are vast, and here I can offer only a cursory and very selective description of some of the main areas of interest. Much has, of course, been left out, and the fields mentioned are vastly more expansive and detailed.

Broadly Marxist theories have been articulated in a number of directions. Early theories, especially those associated with the Institute for Social Research in

Frankfurt, conceptualised the media, and particularly their systematic production within the 'culture industry', as contributory to the imposition of a conservative, anti-revolutionary culture upon a largely passive proletariat (Adorno, 1991; Marcuse, 1991). Another application of Marx's theory regards the media as directly subordinate to the political and economic base structure (Murdock and Golding, 1977; Golding and Murdock, 2000).

The development of a critique of the ideological function of the media brought greater focus on the importance of the media and their influence on society rather than them being regarded simply as subordinate to economic structures. Key in this argument is the work of Althusser (1971), who conceptualised the media as forming part of what he termed the 'ideological state apparatus' within capitalist economies. Here the media were not simply a consequence of the economic system – an 'epiphenomenon' – they were also contributory to its legitimacy. This understanding was problematic for a number of reasons. First, it failed to explain how change and transformation of society could occur; change within a Marxist framework required class conflict. Second, it allowed no 'space' for oppositional political action in the cultural sphere short of a 'total' or Maoist overhaul of the cultural realm. These 'problems' were dealt with in the work of Gramsci, and others who developed his ideas. Gramsci wrote his most significant work between 1929 and 1935 while a prisoner of Mussolini in pre-Second World War Italy. In these *Prison Notebooks* Gramsci (1971) offered an 'autonomous' conception of ideology. Ideology was exercised through the media and wider culture, but it was not something simply imposed upon people; rather, the cultural world functioned as the arena in which contestation and consent manufacture occur (an idea developed in Herman and Chomsky's (1988) popular critique of corporate influence on US media). Gramsci's unrestrictive Marxist approach to ideology proved to be very influential and his ideas were developed in a number of different ways – Hall's (1981; Hall and Jefferson, 1976) theorisation of resistance through the use of media forms, for example. Gramsci's concept of hegemony and consent was also highly significant in the work of radical democratic theorists, as will be examined below.

It has been argued that feminist theories of the media have undergone considerable change in recent years (Press, 2000). Feminist approaches to the media can be understood as being divided into broad schools of thought. In an introductory text Strinarti (1995: 178) argues that three camps or positions, liberal, radical and socialist, can be identified in feminist media theory. Liberal approaches have tended to be concerned with the representational analysis of women in the media, and above all with the 'symbolic annihilation' of women in politically significant public life (Tuchman, 1978). Such analysis of the reflective yet structuring potency of the media is challenged by the work of more radical feminists, who articulate a critique that gender necessitates a greater understanding than merely that of another facet of popular culture (Strinarti, 1995: 189). Instead, Modelski (1986) argues that the very assumptions underpinning notions of mass and popular culture are inherently gender-biased. Femininity is inherently linked to notions of consumption, of reading and passivity, and therefore to an engagement with mass culture. Conversely masculinity is linked to themes of creation, of writing and the production of high culture. The recognition that patriarchal systems of oppression may be integrated with other structures of inequality, including class, gender and race, results in a partial reconciliation of feminist and socialist systems of analysis (Strinarti, 1995: 198–200; van

Zoonen, 1984, 1991). In addition to these approaches, Press (2000) proposes that feminist theory has also concerned itself with three substantive areas in recent years: the renewed interest in the public sphere and women's role in it; the critiques offered by Harraway (1991), among others, of liberal ontological categories of the body, nature and technology; and concerns over difference and identity – the point raised by Felski (1997) that the belief in 'essential' and unifying femininity was challenged, critiqued and derided as a 'meta-narrative' by conservative post-modern theorists. More recently, feminist analysis has also focused upon the so-called 'post-feminist' aspects of media content wherein texts that would have previously been considered misogynist in nature were now considered 'empowering'. Such post-feminist positions have been widely criticised as being little more than deeply conservative views of women thinly veiled as liberation (Levy, 2006; Paul, 2005). Rejecting a notion of 'essence' in the formation of identity and the operation of power, a number of theorists (Butler, 1990, 1993, 1997a, 1997b; Harding, 1998; Kendall and Wickham, 1999) argue for a radical anthropological historical position in which gender is viewed as a historically and socially contingent performance of a set of social power relations.

Such a Foucauldian reading of power resonates strongly with post-colonial theories in which notions of normality are understood as articulated by drawing upon a discourse of ethnic 'otherness'. Similarly, gender theorists explore the socio-historic construction of sexuality through texts. These theoretical positions will be further discussed below in relation to the potential of the radical democratic position.

In relation to the role of the media, the above-noted positions converge upon a general position in which the media are understood to articulate a generally unjust and

99

exclusionary standpoint. The media are broadly believed to articulate, support and proffer systemic patterns of inequality. Far from affording democratic potential, the media are seen as a means by which opportunities are closed off and systems for the legitimisation of inequality are imposed. However, as will be noted below, the completion of such projects is never fully achieved and 'space' for opposition is often found (Torfing, 1999: 223).

The notion of the public sphere as an open arena of discussion is challenged by this critical position. Far from open and inclusive, the public sphere is understood as exclusionary – it holds central a notion of the 'public' that excludes all but a small, white, bourgeois, male, heterosexual élite. The public sphere is far from public, but is comprised only of those who form part of the notional concept of the society in question and excludes all those deemed 'other'. The delimiting of other voices, typically black, female, non-bourgeois and non-heterosexual, indicates, as noted above, an ideological predilection towards the support only of institutionally legitimate political action, a legitimacy that is founded upon normative concepts of citizenship.

Further, as noted above, the notion of the public carries with it an implicit acknowledgement that there exists a separate sphere of activity that is not public, the private sphere. Fraser (1992a, 1992b) notes how certain areas of action are considered 'outside' the public sphere of action, defined in a commentary by Press (2000: 32–3) as 'private issues' and not suitable for public debate. Limiting political action to the public sphere, as liberal democratic perspectives do, seemingly removes legitimacy from political action that does not coincide with that advocated by societal élites and normalised by hegemonic practice.

Unsurprisingly, such an élite is understood to have similar interests with those in control of the media, whose interests are vested in the contemporary capitalist state. Such control is exercised in the maintenance and legitimisation of order that supports the political and economic status quo. However, the nature of the relationship between ownership and control is far from unproblematic and is still a field of considerable debate (Corner, 2000).

Radical democratic

The radical democratic viewpoint refers to a number of positions that incorporate a critical dimension in respect to the liberal democractic notions of the media, but also differ from the orthodox Marxist perspective. Such positions have emerged from the numerous debates occurring in the social sciences in recent years concerning the operation of power within a society. In rejecting political-economic and somewhat neo-functionalist theories of the operation of media in capitalist societies, radical democratic theories accept Foucauldian concepts of the diffusion of power throughout society. Furthermore, Gramscian notions of the deep integration of class and economic interests into cultural forms, prevalent in the work of theorists such as Hall (1981; Hall and Jefferson, 1976), are partially rejected in favour a genealogical account of power formation exemplified by Foucault and theorists from within a radical democractic, post-structural tradition (Laclau, 1990, 2005; Laclau and Mouffe, 1985; Laclau and Zac, 1994). Power is understood as disbursed, decentred and diffuse. It exists not in one central point but in the impact of social systems upon individuals and instances of action. Because of this, resistance to power cannot be targeted at a centralised institution, but must take place at the interface of the individual and culture. Challenging the status quo cannot be performed by big system-level activities; instead, 'local' actions and numerous interventions into the symbolic order may all contribute to, and be considered instances of, political action.

Such resistance is regarded as 'radical democratic action', the rearticulation of discursive moments for the deployment of non-hegemonic identities. Resistance takes numerous forms: for example, Smith (1994) postulates that certain new social movements, already understood by Melucci (1980) as resistance in themselves, are discursive rearticulations of already accepted notions of difference. Radical democratic practice seeks to subvert existing moments of discourse for new, contrary and often political purposes.

Accordingly, the instantiative or totally systemic potential of the media to control the possible permutations of meaning and identity is rejected in favour of a model in which multiple possibilities or readings of media texts are afforded. The media are considered as not entirely 'complete' in their projects of defining political action. Torfing (1999: 223) posits:

> Audiences today may be able to resist the effects of the dominant media configuration and create their own readings and appropriations. The content of messages disseminated by the hegemonic media configuration is only partially fixed. There is always surplus of meaning and a multiplicity of voices which destabilise the dominant meanings and provide material for the articulation of new meanings and alternative political projects.

The role of the media within such accounts is moved away from the position found in many more systemic media theories. While the media may be legitimate forms of oppression and subjugation, there still lies within them the possibility for resistance. The media's role is thus predominately a tool of subjugation, but contains within it the possibility of alternative or radical use. Similarly, the conception of the public sphere moves from one of complete rejection on the grounds of its support for or legitimisation of systemic inequality to one of the possibility of radical action. However, such potential lies in the flexibility of the public sphere, the potential to afford subversion of the existing discourse towards new formations, and not in its existing, implicit, democratic qualities. From within the radical democratic perspective, systemic protection or reservation of the media for citizen usage lies not in the protection from state interference but in protection from commercial interests. Hamlink (1995: 133) notes that while the state may be the perfect tool for the protection of capitalist enterprise through the systematic incorporation of capitalist interest into the legal process, it may also inadvertently incorporate systems of citizen empowerment in the deployment of legal measures. The very legal mechanisms that enshrine capitalist enterprise in Western capitalist states may incorporate the means by which the media may be in part protected from rampant capitalist exploitation. In seeking to preclude monopolist control over media forms, a number of country-specific and international agreements exist that identify non-commercial organisations as the best safeguard. The best protection for the media lies in their being supported by non-partisan public service remits.

The internet as a form of radical media

As Torfing (1999: 223) notes, the failure or lack of complete 'fixture' of meaning in media texts means that they may be used to articulate alternate identities or contrary readings. Thus there is a conception that certain media or the use of certain media may, as Downing (2001) contends, be regarded as 'radical' – media, and uses of media, that consciously articulate oppositional or counter-readings. Radical media emphasise the alternative possibility of meaning in that they consciously stimulate identities and readings contrary to the ideological or hegemonic norm. Such media consciously challenge the legitimate interpretations, readings and uses of media as well as the hegemonic consensus. Downing (ibid.: 8) notes that this is possible, as radical media do not constitute a separate area of cultural life apart from popular culture. They are instead 'part of popular culture and of the overall societal mesh'. Radical media are understood to be present two ways: they work either through the use of existing legitimate media forms with contrary content, popular music and resistance newspapers; or through the use of illegitimate media such as the graffiti and pirate radio mentioned previously (ibid.). Furthermore, radical media may be considered radical in some ways, violating societal norms, but non-radical in others, being conventional and hegemony-supporting – some media forms may be considered as wildly anti-systemic in some regards but at the same time deeply conservative in others (ibid.).

Radical media are understood to afford opportunities for the further exercise of contrary identity formation and the critical reading of cultural forms. More specifically, radical media are understood to provide a means by which alternative discourse can be articulated. Such 'counter-discourse' may fall either within the remit of politically acceptable action or outside it, therefore becoming illegitimate action. Thus radical media as a concept includes media form and use that are both *inside* the legitimate political framework and also *outside* the legitimate political framework. The potential of the internet to challenge the dominant hegemonic forms of media consumption and production, as discussed previously, allows such use of the internet to be considered radical. Indeed, the internet is regarded in this way in a number of instances. For example, Dalhberg and Siapera's (2007) collection explores how the internet may be considered a radical medium through its use in developing various anti-systemic movements such as anti-globalisation, feminist politics, post-colonialism and hacktivism (politically orientated computer subversion).

While various old media do offer opportunities for radical consumption and use, the internet seems distinctly suited for such use. When examined in the light of radical democratic theory the internet, as described in previous chapters, is understood as a radical media form.

Conclusion

The internet is, as noted previously, understood as a media form that specifically affords opportunities for the restoration of democracy or of resistance. It is contended that such notions are situated within either a liberal democratic or a radical democratic framework. It was noted that liberal democratic theories regard the use of the internet as a form of legitimate political action, as the public sphere to which the internet permits fresh entrance has been influenced by, to appropriate Habermas's phrase, 'pathologic' forces. The internet, from such a perspective, allows for a more inclusive form of democratic action.

The radical democratic perspective regards the public sphere and legitimate political action as inherently bourgeois activities. The use of the internet, however, can in certain circumstances be regarded as a 'radical' act as it affords true anti-systemic action, the articulation of contrary identities and the production of media content outside the normal spheres of action.

As indicated earlier, the vast majority of the instances of the usage of the internet examined have been located in either the USA or other late-modern capitalist societies. As such, it is proposed that the internet has tended to be understood solely from the confines of certain Western societies, carrying the 'political baggage' from those societies, the media and the political theory used to explain them. The politics of the West has been used in understanding how the internet functions. As mentioned in the preceding chapter, the internet may well have impacted upon politics, but we must also recognise that our understanding of politics has impacted upon our understanding of the internet.

Furthermore, those studies that have examined internet usage outside the USA and Western Europe have done so from perspectives of late-capitalist societies. These perspectives were developed to explain media usage in late-capitalist countries and have been exported and applied with little regard for localised conditions and preoccupations. The political power of the internet has been understood by examining its use only in certain conditions and/or with theoretical tools that have a particular political form. Accordingly, the internet is understood as operating in a fashion that is implicitly tied to certain conceptions of the political world. The internet is believed to cause change in particular ways, and this is due to the way in which politics is understood.

In the second part of this book I intend to extend this argument to an understanding of how the internet is tied to, and its ability to cause change is dependent upon, society. I will argue that while it may well be true that the internet can affect and change politics, its ability to do so is contingent upon certain social conditions being met. The internet is, in short, contingent.

The next chapter will examine how media and society are deeply linked. I take my lead from a number of sociologists who see the gradual transformation from traditional society through modernity and into a secondary or late-modernity as deeply linked with the emergence of the mass media. Moreover, a number of processes present in and indeed driving modernity, individualisation and detraditionalisation are key to understanding how the internet has been understood and how it can bring about social change. I argue that the power of the internet must be understood as dependent upon and implicitly linked to social processes. The internet may well perform as it has been widely assumed to do, but its ability to do so is deeply tied to social processes and the presence of certain political conditions.

Notes

1   This is an idea popularised by the nineteenth-century Scottish essayist Thomas Carlyle (1859: 147), who argued: 'Burke said there were Three Estates in Parliament; but, in the Reporters' Gallery yonder, there sat a Fourth Estate more important than they all. It is not a figure of speech, or a witty saying; it is a literal fact.'

Part II
The internet and society

6

Social form and media potency – the processes of modernisation

Introduction

The link between a media form, such as newspapers, television, radio and the internet, and the society in which it operates is one of considerable depth and complexity. In the preceding chapters I argued that the way in which the internet has been understood has been influenced by a particular understanding of politics – how politics is practised in certain societies is contributory to how we understand media to operate. Here I wish to develop this argument further: the link between a society and a medium is more 'radical' or deeper than solely a matter of interpretation. While the way in which a medium is understood may owe a lot to how we understand politics to 'function', I argue that the relationship is more complex, deep and bilateral. In this and the following chapter I will argue that the ability of a medium to affect, impact on or change a society in which it is deployed is deeply linked to the cultural motifs and politics of that society.

I will briefly sketch how certain forms of media (the mass media of newspapers, radio and television, and new media, namely the internet) are inextricably linked to particular and historical modes of social life. In arguing for this link I will use a particular model of society and the transformative social forces that occur with modernity. This model is based upon the work of the sociologists Giddens and Beck, who, among others, have made extensive contributions to the sociology of modernity. Indeed, Giddens's work has proven highly influential to the development of much contemporary sociology and how forces of social change are understood (Held and Thompson, 1989: 2). Modernity refers to a social form that has been present in numerous industrial, capitalist societies; it is the 'modes of social life or organisation which emerged in Europe from about the seventeenth century onwards and which became more or less worldwide in their influence' (Giddens, 1990: 1). However, modernity refers to more than cultural life, and this chapter will examine how it is regarded as a system of processes that over time reposition members of society into different social frameworks of meaning.

Of key interest in this area is the proposal that modernity, the nation-state and the mass media are deeply linked. Authors such as Anderson (1991), Schlesinger (1991), Gellner (1964, 1983, 1997) and Thompson (1990, 1995) have examined the integration of the mass media into this association, and their work will be used in this respect.

Furthermore, a number of theorists have proposed that contemporary advanced capitalist societies are now so dramatically different from earlier modern societies that they need to be regarded as post-modern in their social or cultural form. At the very least, the cultural artefacts produced by such late-capitalist societies need to be understood as post-modern, if not entirely symptomatic of a post-modern culture (Waugh, 1992: 3). The former claim is challenged by a number of authors who see the phenomena as symptomatic not of post-modernity but of a later stage in modernity, a social form referred to as 'late' (Giddens,

1990), 'secondary' or 'reflexive' (Beck, 1994, 2007; Beck and Beck-Gernsheim, 2001; Beck et al., 2003; Beck and Lau, 2005) or 'liquid' (Bauman, 2000, 2001, 2002, 2004). This dispute, as to whether the various phenomena represent a break or a continuation, has received considerable attention elsewhere and will not be examined here. Instead, in this and the following chapter I want to focus upon how new social forms, such as post- or late-modernity, entail levels of debate that are best articulated by new media such as the internet. As noted in previous chapters, forms of media enable particular forms of discussion; the mass media afford discussion at a meso, regional or national level, while new media such as the internet allow for a more individualist, micro-discursive mode. Accordingly, this chapter explores the relationship between modernity and mass media before moving on to an examination of the interrelatedness of late-modern social forms and the types of debates that may be articulated through new media.

Defining and understanding modernity

Modernity has been understood as encompassing a variety of phenomena and forms of action. For example, Turner (1990: 6), summarising Weber, proposes that modernity be understood as a social form where:

> the social world comes under the domination of asceticism, secularisation, the universalistic claims of instrumental rationality, the differentiation of the various spheres of the life-world, the bureaucratisation of economic, political and military practices and the growing monetarisation of values.

Moreover, issues are complicated further by mixing terms such as 'modern', 'modernity' and 'modernism'. For example, Habermas (1989: 63–104) refers to the life-world being 'infected by modernism', and 'aesthetic modernity' being equivalent to 'the modern avant-garde spirit'. Seeking to resolve such issues Lash (1987), in a close reading of Weber's work, proposes that the 'modern' be understood by reference to the art movement of modernism, itself a reaction to the challenges presented by modernity. A further problem is interpreting 'modernisation' as a form of progress; modernisation is ascribed values and linked to narratives of development. Undoubtedly some aspects of modernity are progressive. However, as Bauman (1989) has shown, the sheer rationalising power of modernisation can be used to further terrifying ends such as the Holocaust.

Common to most accounts of modernity is the idea that it is a mode or fashion of societal formation that is in some way distinct from what went before. Modernity is a specific form of social life that is in many ways radically different from previous forms of social life. It is perhaps most clearly defined by its conscious emphasis upon opposition to, or movement away from, tradition. Modernity is explicitly articulated as a form of social life that is based upon a rejection of the 'old'. However, as Giddens (1994: 56) notes, in many instances modernity has 'rebuilt tradition as it has dissolved it'; tradition serves in many Western modern societies as the basis and legitimisation of political power.

Giddens (1991) further argues that modernity and the formulating processes of modernisation are complex and multidimensional. Modernity emerges in relation to, and is characterised by, three interrelated modernising processes: the separation of time and space, the disembedding of social systems through abstract systems and the reflexive ordering and reordering of social life. It is worth addressing each of these in turn.

In pre-modern systems, time and space were to a large part deeply integrated. Giddens argues that virtually all social systems have a way of ordering time, but in pre-modern societies time and place were deeply linked. Time was understood as tied to socio-spatial markers – it was not possible to understand time outside the spatial location in which a referring event occurred. Within modernity, time-space linkages were shattered; time becomes, to a certain degree, universal and conceivable away from spatial location. Giddens proposes that the significance of such a separation is threefold. Firstly, the separation of time and space leads to the possibility of 'disembedding', which is discussed below. It allows modern life to 'break away' from local practices and habits. Secondly, rationalised organisation is made possible by the disjuncture of time from space. While authors such as Weber noted the static and repressive nature of bureaucracies, the 'iron cage of bureaucracy', Giddens notes their dynamism when compared with pre-modern social systems. Modern bureaucratic systems afford a connection between different levels of social life, between the 'local and the global', ideas that were simply unthinkable in pre-modern times. Thirdly, the disjuncture of time from space affords the possibility of the appropriation of history in a unified manner that will allow for planning for future events.

Giddens (1990: 22) argues that the process of disembedding social systems distinguishes modernity from pre-modernity. Dissembedding refers to the way in which social change is related to the 'shifting alignments of time and space'. Giddens identifies two types of disembedding mechanisms present in the shift from pre-modernity to modernity. First is the development of symbolic tokens, of

115

which money is considered to be the most significant. Money provides a way in which space can be obtained between individuals and their possessions, emphasising time-space distanciation. Modern economic systems extend this disembeddedness to a greater degree than in any pre-modern system, and money accentuates the degree of time-space dislocations to a more acute state. Giddens argues, however, that money does not 'flow' but allows the opportunity for instantaneous transfers of power that were inconceivable under pre-modern systems. Secondly, Giddens (1994: 21) draws attention to the multiplicity of 'expert systems' – the related and regulated procedures that permeate modern life. Such systems provide methods of control for the multiple forms of activity that occur in modern life; examples include traffic systems, health and safety conventions and any other corpus of knowledge disembodied from its originating discourse that affects social life.

Giddens (1990: 53) notes that all human action contains a degree of reflexivity – even traditional societies do not unquestioningly follow the prescriptions of previous generations. Anthropological studies illustrate how traditional societies are aware of what their traditions do and the possible functions that they fulfil. What differentiates reflexivity in the age of modernity from previous systems is the way in which 'the production of systematic knowledge about social life becomes integral to system reproduction, rolling life away from the fixities of tradition'. Within modernity there has been a systemic appropriation of the knowledge of society for the legitimisation of social action. This is substantively different from pre-modern societies, where traditions legitimated action. Tradition may still be used in the legitimisation of action within modernity but it is no longer the final point, rather tradition itself is legitimated by reflexive action and by appropriation into rationalised action.

It is these three processes, the separation of time and space, the disembedding of social systems through abstract systems and the reflexive ordering and reordering of social life, that have resulted in the specific social form of modernity in the West. Integral to such a social form are the institutional entities integrated acutely into many facets of social life. Giddens views these as operating in four distinct but deeply interrelated dimensions of life: capitalism, surveillance, military power and industrialism.

We can define these terms further. Capitalism describes a mode of distribution centred upon private ownership of the means of production. Ideally, it involves a specific separation or 'insulation of the economic from the political' (ibid.: 59). While capitalism itself tends to be a transnational phenomenon, it requires systems that can only be instantiated within the remit of nation-states. Surveillance refers to the 'supervision of subject populations in the political sphere' (ibid.). It involves not only the direct application of power in a Foucauldian sense, but also the control of information. Surveillance technologies and procedures necessary for the administration and deployment of contemporary capitalism could only exist in modern societal forms. Military power involves the monopoly of violence and control over the means of institutional violence, such as police and armed forces, within the boundaries of a nation-state. While violence was a constant feature of pre-modern societies, such concentrations tended to be short-lived. Modernity sees the centring of political projects within the legitimate state apparatus, affording long-term, and to some degree peaceful, transitions of power. Finally, industrialism refers to the broad system of means of production that involve the 'inanimate sources of material power in the production of goods' (ibid.: 55–6).

These institutional dimensions operate across national borders. Indeed, Giddens (ibid.: 64) notes that sociology, in its attempts to understand modernity, may be better off concerning itself with 'how social life is ordered across time and space' than just within national boundaries. Globalisation is widely understood to be a force detrimental to the existence of nation-states. For example, Appadurai (1990) has argued that the new flows of people, technology, ideology, money and culture will lead to a gradual decline in the power of the nation-state. Giddens (1990: 64) challenges such views by proposing that the legitimisation of nation-states arises not from some historic form but from being in an international system: 'the sovereignty of the modern state was from the first *dependent upon the relations between states*' (emphasis in original). The modern world system was international from the outset – it was always a globalising force. Modern life has not been about the decline of the nation-state. Instead, Giddens (ibid.: 67) argues that:

> The history of the past two centuries is thus not one of the progressive loss of sovereignty on the part of the nation state… Loss of autonomy on the part of some states or groups of states has often gone along with an *increase* in that of others. (Emphasis in original.)

Indeed, as Castells (1996) contends, modern social systems with their wide-scale bureaucracies, rational systems of law, conscious denouncement of traditional knowledge in favour of rational bodies of knowledge (while in many instances still drawing upon tradition for legitimisation), industrial systems of production and capitalist mode of distribution were exemplified by nation-states. The nation-state provided a means by which the systems of modernisation, as detailed by Giddens, may be instantiated into a form that could be coherently proffered to citizens or subjects. If the nation-state provides a system by which the processes of modernisation may take place, there is also a degree of reciprocity in that the nation-state is understood as a consequence of modernity.

Modernity or modernities?

In recent years an interesting discussion has taken place that examines the idea that the categories Giddens and others use so freely to define modernity, such as capitalism, may not be single entities. Indeed, there is a new body of literature emerging from political economy that examines varieties of capitalism and asserts that it should be regarded not as a singular but as multiple entities (Hall and Soskice, 2001; Yamamura and Streeck, 2003). There are differences between the various capitalist systems, but the authors state that the differences between the German model and the Japanese model need to be considered as differences in *mode* rather than essential differences in nature. The different instances should be considered as differences in a *family* of economic formations. Thus they need to be considered as varieties of capitalism rather than different capitalisms.

This idea has raised an interesting challenge to one major critique of Giddens: that his model is based upon and accounts only for European experiences of modernity (Burawoy, 2000; Harindranath, 2006). Schmidt (2006) argues that rather than there being multiple modernities, as many authors argue (see for example Kaya, 2004), there are instead varieties of a common set of practices, or varieties of modernity. Thus modernity in various locations may not be exactly the same, but it will have a core of similar processes and characteristics that liken it to other varieties of modernity.

Modernity and the nation-state

In examining the relationship of the nation-state to modernity it is fair to say that sociology has perhaps subjected the study of the nation-state to 'benign neglect' (Thompson and Fevre, 2001: 297). Of the accounts that do exist, sociological or contextual studies – those that emphasise the historicity of nations and challenge the 'primordialist' conception that regards nations as emerging out of instinctual drives (Kellas, 1991: 36) – prove particularly useful in this context. Here I will focus upon the somewhat complementary theories of Gellner (1964, 1983, 1997) and Anderson (1991). As indicated previously, modernity and the nation-state are deeply linked. However, the nature of this linkage is complex and multifaceted. Gellner's proposal, first set forth in 1964, elaborated in 1983 and restated in 1997, contends that the nation-state and nationalism emerge out of the conditions of modernity. Moving away from approaches, such as Kedourie (1993), that regard the nation-state simply as a by-product of the progress of modernity, Gellner (1997: 11) argues that the nation-state '*is* the necessary... correlate of certain social conditions, and these do happen to be *our* conditions, and they are also very widespread, deep and pervasive' (emphasis in original). Gellner (ibid.: 13) proposes that modernity offered the correct conditions for the emergence of nationalism and even caused it: 'nationalism is rooted in modernity'. He views the relationship between nation-states, nationalism and modernity as inherently reciprocal.

Gellner's model of the emergence of nation-states describes how societies change in relation to industrial development. Gellner proposes that nation-states emerged as the most appropriate way to deal with the demands of industrialism. Antique social conglomerations were modified and were in turn reconciled to the necessities of industrial production. Attendant to the emergence of a nation-state as a political entity was the necessity of nationalism, the articulation of discourse that allowed the nation to be imagined by its members. Focusing upon central Europe in the post-Napoleonic Wars era, Gellner describes an abstract model of the way the 'nation' emerges through a series of five transitions or stages in societal consciousness.

This is a theme also followed by Anderson (1991), but while Gellner is primarily concerned with the emergence of nations as political entities, Anderson is concerned with evolution of the 'psychic' qualities of nations and nationalism. Particular attention is paid towards how, in addition to the material conditions that allowed for (and perhaps necessitated) their emergence, nations also needed to be 'imagined' – 'What makes people love and die for nations, as well as hate and kill in their name?' (ibid.). Anderson contends that it was a convergence of certain conditions, occurring in early modernity, which afforded the emergence of the 'imagined community' of the nation. Anderson proposes that the emergence of print technologies along with the rapid growth of mercantile capitalism resulted in the gradual decline of the variety of languages spoken in a region. This afforded the potential for the emergence of a 'national print language' that consequentially allowed for the 'possibility of a new form of imagined community, which in its basic morphology set the stage for the modern nation' (ibid.: 46). Furthermore, Anderson explores the way in which the nation allows for emotional commitment by citizens. This affords a continuation or replacement for the feelings of belonging that industrialism was removing from traditional systems of affiliation, such as the extended family group or clan. The processes of modernisation that brought about new systems of production consequentially afforded new means of affiliation between individuals. The articulation of such 'imagined communities' of nations was possible due to the extension of print technology and the widespread understanding of a national language.

Modernity and the mass media

Mass media are often understood to be an essential element of modernity and the emergence of nationalism. Gellner (1983: 127) proposes that 'it is the media themselves, the persuasiveness and importance of abstract, centralised, one to many communication, which itself automatically engenders the core idea of nationalism'. It could be argued, however, as Schlesinger (1991) does, that Gellner regards the mass media and the resultant 'national identity' as little more than by-products of the formation of a nation-state, which is itself a consequence of the technological innovations of the eighteenth, nineteenth and twentieth centuries. Furthermore, Gellner's theory seemingly lacks a means by which national culture is actively invented and recreated. It is without the Andersonian dimension of a social, spiritual or cultural construct of the nation.

Thompson (1990, 1995) offers a solution to this problem by providing an extensive model of the deep integration of the mass media with modernity. For Thompson the emergence of the mass media involved the development of technical and institutional means of production and diffusion – the development of technical innovations that were commercially exploitable, and exploiting them in an institutional fashion. He argues that mass media arose in part because of the institutional adoption of certain technical methods and technologies – institutions that were only possible with the forms of systemisation that occurred in modernity. It is in the development of such institutional systems that Thompson identifies a considerable degree of interrelatedness with mass-communication systems. Thompson (1990: 218) proposes that the mass communication that typifies accounts of old media can be distinguished from earlier, more individual-to-individual communication by its intention to be available 'in principle to a plurality of recipients'. In recognising this, Thompson acknowledges that while certain old media forms are consumed by large numbers of people, it is not the actual number of consumers that distinguishes mass media but rather the principle that they may be available to large numbers. Furthermore, mass communication is primarily a non-dialogical form of communication: information flows in only one direction, typically from producer to consumer. With these two main caveats Thompson defines mass communication as 'the institutionalised production and generalised diffusion of symbolic goods via the transmission and storage of information/communication' (ibid.: 219).

Moreover, the mass media made possible the very idea of a national culture that Anderson (1991) identifies as vital to the development of the nation-state:

> The emergence of a sense of national identity – and indeed of national*ism*, understood as the channelling of national identity into the explicit pursuit of political objectives – was closely linked to the development of new means of communication which enabled symbols

and ideas to be expressed and diffused in a common language. (Thompson, 1995: 51, emphasis in original.)

Furthermore, Thompson (ibid.: 62) regards it as highly plausible that:

> The formation of national communities, and of the distinctly modern sense of belonging to a particular, territorially based nation, was linked to the development of new systems of communication which enabled individuals to share symbols and beliefs expressed in a common language – that is, to share what might roughly be called a national tradition – even though these individuals may never have interacted directly.

This is an argument earlier proposed by Scannell (1988: 29), who argues that the mass media and specifically broadcasting:

> unobtrusively restored (or perhaps created for the first time) the possibility of a knowable world, a world-in-common, for whole populations... broadcasting brought together for a radically new kind of *general* public the elements of a culture-in-common (national and transnational) for all.

The mass media, an institutional entity built into and articulating the economic and social fabric of the nation, affording symbolic forms specifically produced for public consumption outside the temporal and spatial location of their production, allowed for the construction of national identities. Mass media operate in concert with other features of modernity and help to express and formulate the modern form of the nation-state. Modernity is implicitly defined by its pre-eminent form of social affiliation, the nation-state – an entity that is only possible with the emergence of the mass media. Modernity exists in an intricate relationship with mass media: mass media afford the expression of those forms of identity that typify modernity, while modernity provides the social forms and necessary institutional frameworks for the mass media to exist.

Late- or reflexive-modernity and the individualisation of society

The majority of the authors drawn upon here, Giddens, Beck and Thompson, and in earlier chapters Slevin, all share a common theoretical standpoint – indeed, both Thompson and Slevin were students of Giddens and their work is highly influenced by him. These authors broadly concur that the changes so widely noted in academic circles and understood as symptomatic of a post-modern society are, in actuality, an acceleration or extension of the processes of modernity. Contemporary society is best understood not as a period beyond modernity, but as a radicalisation and continuation of the preoccupations and systems of modernity.

Regarding certain societies as experiencing late-, high- or reflexive-modernity is not an acceptance that modernity has ended, rather it is an assertion that modernity has entered a new phase. While late-modernity may appear radically different from modernity, it is actually a continuation and extension of the central concerns of modernity – an acceleration of features and processes that have been present throughout modernity.

Individuals in late-modern societies experience a number of transformative forces; perhaps key among these are the twin pressures of individualisation and detraditionalisation. Individualisation, a term that has gained increasing currency in recent years, refers to a general transformation in the mechanisms of identity formation away from collectivity and towards individualism. Giddens (1991: 32) describes how the processes of modernity, particularly the greater reflexivity or constant emphasis upon the 'self' as a project, bear upon the individual to such a degree that 'the self becomes a *reflexive project*' (emphasis in original). Individuals are no longer grounded in the virtually unchanging systems of tradition (ibid.: 32–3).

These ideas have been developed by Beck and a number of others. Beck initially sketched out a model of individualisation in his 1992 text *Risk Society* and then more fully elaborated upon the idea in a number of other texts (Beck, 1994, 2007; Beck and Beck-Gernsheim, 2002, 2004; Beck et al., 2003; Beck and Lau, 2005). For Beck, individualisation is a process, a gradual transformation of the way in which individuals are able to experience life. Importantly, individualisation is not a conscious choice made by people to live their lives in a particular manner, rather it is a macro-sociological transformation – a consequence of the destructuring and restructuring of class and of other changes in advanced capitalist societies (Beck, 2007: 680). Such changes occur as the background to security and stability provided in modernity by the state; other agencies are, in reflexive-modernity, devolved to the individual. Beck refers to this process as 'insourcing', the devolving of responsibility back upon the citizen. Similarly, in terms of emotional activity, individualisation impacts as family and divorce laws conceptualise the subject as 'individual' to a greater degree than previously (Beck and

Beck-Gernsheim, 1995, 2004). Individualisation is thus present in the changing nature of the relationship between the individual and the state. It is a trend that has been present throughout modernity, but is greatly accelerated in reflexive-modernity. In contemporary society it is most acutely felt when it is 'imposed by modern institutions' (Beck, 2007: 681).

Linked closely to individualisation is the detraditionalising force of modernity. As noted earlier, modernity is by definition a movement away from what went before, an attempt to move beyond historical pre-modern systems of thought and action. Heelas (1996: 2) defines the detraditionalisation process as resulting in 'the decline of the belief in pre-given or natural order of things'. It is the forward-facing, conscious effort to progress and distance ourselves from preceding views, to move beyond ancestral world views. While such overtly 'progress'-oriented attitudes are an inherent part of modernity, they often constitute a 'straw man' against which 'quests for authenticity' such as nationalist movements articulate themselves. Indeed, detraditionalisation may be most strongly felt and witnessed by the attempts to challenge it, to quell the fear of losing the 'traditional' in the face of the 'modern'.

Giddens contends that such processes are peculiar to societies encountering high-modernity. For Beck and Beck-Gernsheim (1995: 24), individualisation in Western liberal democracies refers to the '*dis*integration of previously existing social forms – for example the increasing fragility of such categories as class and social status, gender roles, family, neighbourhood, etc.'. This decline of the potency of existing systems of society to define action does not mean, however, that the individual is now free from social systems. Rather, in late-modern societies, 'new demands, controls and constraints are being imposed on individuals' (ibid.: 25). Individualisation is not the liberation of the individual from social regulation, but, with regard to the regulations and structures of life, 'individuals must, in part, supply them for themselves, import them into their biographies through their own actions' (ibid.). The potent features of life, the systems by which meaning is made and identity formed, are now available in a multitude of ways from which the individual now chooses. The waves of modernisation, the processes that Giddens describes, result in a sequestration of the individual from the fixed systems of early modernity. Individualisation is the forward-pointing process that arises as a consequence from the radicalised extension of modernisation. It is not simply a disintegration of old social forces, but the new system of subject formation, a system that affords or demands greater individual responsibility.

The internet as a technology of individualism

In earlier chapters I sought to identify a predominant conception or understanding of the internet. I argued that the internet is widely seen as a technology that frees the individual from the fixities of identity formation understood to take place in older forms of media. The internet is a medium thought to give individuals the opportunity to choose their own path through media and define or 'weave' their identities. These opportunities, and the activities they afford, bear a strong resemblance to the description of individualised identity formation. I want to argue that this resemblance is significantly more than a coincidence, for a number of reasons.

First, that such qualities have been ascribed to the internet is due to the site of the emergence of technology and the predominant cultural ethos of the manner in which the internet has been understood. As discussed in the preceding chapters, a dominant understanding of politics has played a considerable part in how the internet has been conceptualised as a medium. The internet was developed and researched in societies that were considerably individualised. It was understood to afford such qualities and preoccupations as were present in these societies. It has been viewed through a 'template' of the values and interpretations of the societies in which it was developed and most often studied. Therefore, as the internet has been envisaged through the values, cultural machinery and explanatory frameworks of high-modernity, it will of course display the very values and preoccupations, such as a concern with identity, that these societies use to make sense of the world.

Second, at the same time the media cannot be completely divorced from society and regarded as passive. As examined above, Thompson explores the deep links between modernity, the nation-state and the mass media. I wish to argue that the internet should be viewed in a similar manner, and that the internet and life in late-modernity are deeply linked. Poster (1997: 80), describing how the internet affords new avenues of identity formation, notes: 'electronic communications technologies significantly enhance... post-modern possibilities'. The internet offers specific opportunities for communication that were not possible with older forms of media. It makes possible precisely the identity-forming actions associated with individualisation. Turkle (1995: 180), following an argument about the possibility of playing with identity on the internet, proposes that internet technologies are a 'social laboratory for experimenting with the construction and reconstruction of the self that characterises post-modern life'. The internet articulates the social forms and debates of late-modern societies – precisely the conditions in which the internet emerged, where it is mostly used (at least until very recently, with the virtually exponential growth in user numbers in China in the past few years), where it has been most often studied and, perhaps most importantly, where the originating discourse of the internet as a medium of individuality emerged. The internet is deeply linked to the social form of late-modernity – it affords the activities and articulation of the debates that typify and define late-modern societies. It makes late-modernity possible in a manner no previous media form could. Media such as the internet are an essential part of late-modern life – these are systems that intersect with lived experience on a daily basis. They cannot be regarded as external to lived life, rather they are a part of life in late-modernity; they make possible, afford and allow the late-modern way of life. As Lash (2001: xiii) argues:

> It is at the interface of the social and the technical that we find the second-modernity's individual. It is at this interface that we take on the precarious freedom of a 'life of own'; that we 'invent the political'... the individual in the second modernity is profoundly a socio-technical subject.

The internet is not only interpreted through the 'template' of the cultural patterns of late-modern society, it also contributes to and makes possible these very patterns. The link between the internet and society is not simply one of technology causing a social form, or of a technology arising out of a social form. Instead it is one of deep integration;

technology is both understood through cultural patterns and also affords those cultural patterns.

Conclusion

In this chapter I have sought to explore how the relationships of old media with modernity, the nation-state and the attendant forms of political identity that arose are deeply linked and cannot be separated. Mass media did not cause the occurrence of modernity, nor were they incidental and consequential of modernity; rather they are best regarded as deeply linked in a complex bilateral manner. Modernity arose in concert with, but at the same time made possible, systems of mass media. Furthermore I sought to examine the argument that modernity is not concluded, but that it may best be viewed as a set of processes that are still active in the integration of the individual and society. Late-modernity is understood to be a period of radical individualism. It is a form of social life in which macro-sociological processes necessitate that individuals 'build' their own identities. Such a 'preoccupation' with or understanding of the social world finds strong resonance in discourse surrounding the internet. I have argued that this is not surprising for two reasons. Firstly, the internet arose in societies in which such preoccupations prevail. In such societies, the internet is the medium that offers the level of debate that is of key interest. Secondly, technologies such as the internet are part of the manner in which late-modern life is actually experienced and lived. They make possible the expression of identity that characterises late-modernity.

In the following chapter I will sketch a model for understanding how we can begin to comprehend the internet not as a technology that either causes social change or is entirely subservient to social action. Rather, the approach I will advocate is one in which the internet is conceptualised as having the potential to cause social change, but that this ability is contingent upon social conditions. The internet can bring about social change but is not external to social forces, rather it is imagined through, understood with and contributory to social and cultural patterns.

7

The internet and society: reconsidering the link

Introduction

In this chapter I want to bring together ideas that have been introduced in earlier chapters and offer a 'model' for understanding how the internet impacts upon society. In brief, the proposed model is a third position between the arguments of McLuhan and Williams outlined at the start of Chapter 2. As was noted then, these positions have been widely influential in how the internet has been understood to interact with society. Central to the alternative third position proposed here are two key arguments.

Firstly, the media should be viewed neither as *causing* social change, as someone sympathetic to a broadly McLuhanist position would suggest, nor as entirely *subservient* to the 'social', as someone aligned to a broadly Williamsonian position would argue. Instead I propose that the relationship between media and society is far more complex, and its understanding needs to be more nuanced. Media technology, such as the internet, is a social phenomenon: it should be considered part of the social world, not external to it. Moreover, it is imagined through social and cultural values or 'templates'. In a complex act of reflexive interpretation the values of late-modern society are used to understand how the internet impacts upon such late-modern society. Partly because of this I argue that media can and do contribute to how we live and experience the world. However, the *way* in which media interact with society is not one of simple 'cause and effect'. Instead, the argument put forward here is that media technologies and forms such as the internet *facilitate* and *accelerate* the social behaviour, interpretative frameworks and mid-level social patterns and systems that dominate and characterise late-modern life. In brief, the internet does not cause late-modern life, but it does make it possible and accelerates the processes that underpin it.

Secondly, expanding upon this point I argue that the internet needs to be considered as a 'contingent' medium. The internet can and does do as so often described: it can contribute to changes in society and individual lives. It does make possible the manner of living in late-modernity, and all that this entails in terms of personal politics and the interpretative frameworks used. However, for such activities to happen the conditions have to be correct. If we see media technology as part of society then we must appreciate that those conditions that make it possible to function are not incidental to its success, rather they are core to it. Therefore to understand the use of the internet in a particular situation we must look at the social environment as well as the purely technological aspects. Methodologically, the internet should not be the only field of analysis. Focusing singly upon technology without examining the social situation of its use leaves unexamined a major factor in the internet's operation. As Miller and Slater (2000: 11) argue: 'if you want to get to the internet, don't start from there'. Thus to study the internet successfully we need to look at its use in relation to social systems and not as an external, discrete contributor to them.

One result of accepting these two propositions is a more socially oriented account of the internet. It is a position in which the internet is regarded as a component of society rather than external to it. Furthermore, regardless of its real power or 'materiality', in this interpretation the internet shifts from being a discrete, objective entity to a socially imagined one. Considering it as a social entity will allow us to apply to it the same explanatory devices we use for investigating other social phenomena. It becomes an entity that we can explore with sociological tools as we do with other social phenomena of the late-modern social world, be they social practices such as the family or the institutional interpretation of such entities through regulatory and institutional procedures. We can further start to explore the relationship between the internet and these other social phenomena in a meaningful manner without resorting to describing the internet as more 'real' than other areas of the social world.

A second result of these ideas is that we can or should recognise the limitations of the internet to advance progressive ideals if we fail to match technological development with social and political development. Unfortunately the internet, as with similar technologies previously, has been regarded as a 'quick fix' for social problems. The internet, being part of society, cannot alone initiate development or bring about progressive social change; it can only do so in partnership with political and social will.

The argument so far...

In Chapter 1 I proposed that a number of broad methodological standpoints exist in the study of the internet.

The first standpoint, which dominates technological fields, tends to focus entirely upon technology to the detriment of all other modes of analysis. This approach ignores all notions of the user beyond biological capabilities and cognitive skills. The second approach reverses this position, and seemingly ignores the medium itself and examines only the nature of the communication that takes place. Technology is demoted and reduced in importance in terms of analysis. Here, attention is focused upon notions of 'text' and of the motivation of interlocutors, either social or psychological. The third approach reintegrates technology and studies the impact of technology on communication. Technology is 'raised' in terms of its significance within this model of communication. This approach examines how the qualities of the media affect the form of communication. It is this model that informs much contemporary comment on the internet from within the social sciences and humanities.

It was also noted that the idea or definition of the internet has, in many ways, been left unexamined. The internet is regarded as a phenomenon existing outside society. It was proposed that there is an emerging, alternative, sociological perspective. This position, while accepting a general notion of the potency of technology, challenged uncritical and discrete accounts of the internet and society. This book was written to elaborate and extend this position.

In developing this methodological stance, the book is divided into two parts. The first part consists of a multilevel description of contemporary understandings of the internet. Chapter 2 detailed a number of different ways in which the relationship between society and technology has been thought of. As with Winston (1998), it was argued that technology should be understood in a Kuhnian fashion – that it is a manifestation of the system of science of the time. Technology does not possess any degree of 'universal truth';

rather, it articulates a set of scientific beliefs. It was further argued that historically there have been three distinct interpretations of the way in which technology impacts upon society: technological instrumentalism, technological determinism and technological substantivism. Such interpretations continue to inform how we see technology and, perhaps more importantly, these positions are very evident in discussions surrounding the internet.

In Chapter 3 I sought to describe the main ideas surrounding the operation and features of the internet. This described the way in which the internet has been broadly conceptualised in texts from journalism, the social sciences and the humanities. I argued that within such literature a general perception exists that the internet is different from previous media forms. Such difference is understood to result in the user being afforded alternative and significant forms of action. I proposed that the internet is generally understood to possess four qualities that distinguish it from older forms of media: human-computer interactivity (the ability of a user to control the flow of information), geographically distributed users being able to communicate using the internet, the production and dissemination of content by users and the personalising of media to individual tastes and needs.

In Chapter 4 I examined how the internet has been understood as a means of communication that may challenge the existing forms of political action. This challenge is regarded as possible because the internet is viewed as possessing qualities unavailable in previous existing media forms. Literature in this area is dominated by debates concerning Habermas's concept of the public sphere and the notions that the internet can bring about social change. The internet is widely assumed to afford action that will allow for the re-establishment of the public sphere and the deployment of new forms of identity that lie outside the hegemonic mainstream. The internet has been conceptualised as an inherently democratic medium; it has even been conceived of as a medium that affords opportunities for radical new action.

The way in which this contention was consistent with a wider model of the media, politics and democracy was explored in Chapter 5. The concepts of the internet as a restorative agent, able to redress imbalances in democratic practice or to offer new and radical means of communication, were argued to be politically situated assertions, located within specific Western political frames of reference. It was argued that the internet had been conceptualised through two broad schools of thought (though with many minor variations). The first school is composed of those theories that regarded it from a 'liberal democratic' viewpoint. In this conception, the internet is regarded as a 'small' medium – a media form that, while legitimate, could at least partly exist outside the corporate or state media systems. It was viewed as a means by which individuals who had been excluded from the public sphere could re-enter it and have their voices heard. The second school, termed 'radical democractic', incorporates a number of criticisms mounted against the liberal conception of the media. From this conception, the internet was viewed as a form of 'radical' media. It focused upon the internet's radical potential and its ability to afford truly anti-systemic discourse.

The underlying rationale of this multilevel description was twofold. Firstly, it was to document the beliefs surrounding the internet. Secondly, it was to demonstrate that the ideas surrounding the internet, particularly its interpretation as having the potential to affect politics, were firmly grounded in particular models of politics and social life. This idea, that the existing models of the internet's political role were grounded in a set of social and political beliefs, was foreshadowed in Chapter 2 when three models of technology's interaction with society were described. The placing of this chapter at the start of the book was intended to demonstrate how even fundamental perceptions of the potency and description of technology are not certain, essential or eternal. Rather, they are firmly rooted in a particular social formation, that of modernity. This is key to the second aim of the book: to offer a methodological standpoint for the study of the internet.

In Chapter 6, which commenced the second part of the book that will be concluded in this chapter, a model of the relationship between media form and social form was proposed. It was argued that mass media are deeply interwoven with modernity, the nation-state and certain forms of identity. I argued that mass media do not cause the emergence of such social phenomena, but are intrinsically linked to them. It was further contended that societies with a late or high form of modernity constitute identity in a different fashion to societies in modernity. High- or late-modernity is understood as involving a radicalisation of certain trends in modernity – the processes of detraditionalisation and individualisation. In such societies identity is constituted in a highly individualised fashion, a trend that seems strongly echoed in discourse surrounding the internet.

It was proposed that the internet has been regarded as a technology that carries with it a social-determining power; that the internet may cause social formations to arise. This was regarded as problematic; it is a seeming rearticulation of the technological substantivist argument that technologies carry with them a stain of their situation of manufacture, and through their use structure action around them. The current understanding of the internet, particularly of its political potency, involves a reification of political values on to the technology. As noted above, such an argument arises because the societies primarily from and in which the internet has been studied are those where the very potencies that the internet is assumed to afford are social currency. I propose that this is incorrect; instead, the internet should be regarded as a *facilitator* and *accelerant* of social process, not as a *cause* of them nor as a *servant* of them.

Social systems and the active nature of the internet

Where I noted in Chapter 3 that the internet has been generally conceptualised as possessing certain characteristics, here I want to offer my own characterisation. However, where in Chapter 3 I discussed how the technology was regarded as possessing unique qualities, here I argue that it intersects with social processes. The internet needs to be considered as an *active* and *contributory* part of the social world. By active and contributory I mean that it is not merely a passive element of social life. Rather, the internet is active in shaping daily life and lived experience. The internet operates at the 'mid-level' of social life; it acts upon the social systems and forms of collaborative understanding that make daily life possible.

For sociologists such as Giddens and Beck social systems are a model for understanding how everyday life is ordered and made possible. Sociology has often been characterised as having two differing traditions. There are those approaches that seek to identify the meta-rules of social life, the social structures such as class and gender. These approaches have sought to examine how and why society is as it is, to determine the rules that underpin macro-level or historic changes or to identify the 'function' of certain social phenomena. Society is regarded as broadly similar to the natural world, and the same methods that are used in natural science can approximately be used to identify the 'hidden' reality or meta-laws of social life. A parallel approach challenges this positivist tradition and argues that attention should be focused upon the micro-rules that govern behaviour and the decisions that individuals make in their daily lives. In such phenomenological and ethnographic approaches it is the decisions of individual actors and the interpretation of their actions that should be studied.

The concept of social systems[1] provides a means by which these two approaches can be at least partially reconciled. In an account proposed by Giddens (1976, 1979, 1984), social systems are identified as the everyday rules and routines that we use to live by. They are the social practices that dictate how we act in particular situations and how to perform activities. In turn they provide the 'language' that allows us to live in society, the means by which we can plan for and understand our world. Social systems provide us with routines to live our lives, the necessary forms of collective understanding that we need to communicate and cooperate. They 'insulate' us from the minutiae of moment-to-moment life; because of social systems we know what to say when some one greets us, and how to function in particular situations. Social systems provide frameworks through which we can exist and live in a social world. However, present in social systems are larger social structures. Social systems are regarded as the 'media' through which social structures act upon us. While social structures cannot be directly viewed or measured and can be known only through their effects, they are made evident through actual social practice – they exist and impact upon us through social systems. Furthermore, social systems must be understood as learnt and actively reproduced codes of behaviour. We learn the conventions of correct behaviour, of living in particular social settings and of communicating in a certain manner in certain conditions. However, we also reproduce and contribute to their continuity through the performance of the social systems. Because of this constant reproduction, social systems are not static but are subject to change and transformation. In the contemporary era, the nature and direction of this change and manner of the transformation may be understood by reference to the transition in social form from modernity to late-modernity, as discussed in the preceding chapter. Social systems are thus a means by or a site in which social transformation can be witnessed. The gradual changing of social practice, be it the socially accepted guidelines for conducting romance, the formality of communication between academic and student or an individual's interaction with state and organisational bureaucracies, serves to demonstrate the transition from modernity to late-modernity – it is in social systems that we can identify the processes of individualisation and detraditionalisation that characterise late-modernity.

The internet has intersected with innumerable social systems of late-modern life and is itself comprised of such systems. It subtly intersects with and adjusts such systems, and plays a part in their transformation. The specific manner in which it does this occurs in two distinct ways: the internet *facilitates* and *accelerates* the particular social processes that dominate and characterise late-modern life and inflect social systems.

The internet as facilitator

By facilitation I refer to the manner in which the internet allows characteristics of late-modernity to be experienced. We can regard it as a conduit through which the power and dynamic nature of late-modernity are disseminated throughout society and allow society to *be* late-modern. The internet is a means by which late-modern social practices are subtly disseminated into social systems. This occurs in two ways.

First, it allows greater reach for the processes, an increase in the *breadth* of their impact upon social systems. The internet results in a seemingly ever-increasing number of people engaging in those activities that characterise late-modern life. As the numbers of people using the internet continue to rise, so the processes of late-modernity, individualisation and detraditionalisation impact upon more and more people and the social systems in which they partake. However, two points need to be made here. Usage is not consistent either across the developed world or within countries. Both individual countries and regions such as Europe have significant levels of disparity in the percentage of the population being online. Also, there are considerable differentials of usage between demographic groups, with youth and education being the most significant correlates of internet usage. Such instances of the 'digital divide', while persistent, are gradually decreasing, but there do remain pockets of 'hard-to-reach' groups. It must be remembered, however, that at the same time as offering new opportunities, extension also bring other aspects of late-modernity such as increasing surveillance and the quantification of everyday experience ever closer to the individual. These aspects should not be regarded as 'good' or 'bad', though of course they may have positive or negative consequences for an individual, but instead must be understood as part of late-modernity.

Second, the internet allows a greater *depth* of experience of late-modern social processes. Individualisation, while a constant feature of late-modernity, is brought to the attention of individuals to a far greater degree when they are confronted with the innumerable potentials of individualism and individual identity formation present in internet communications. While individualism and individualisation are distinct (Beck, 2007), the assertion of individuality, the conscious sequestration of the individual from the group, is a symptom of the underlying drive for individualisation occurring in late-modernity. The conscious decisions we must constantly make mean that communication using the internet brings to the fore, and makes evident, the continual demand to define ourselves – a requirement that is a direct consequence, and key to the process, of individualisation.

The internet as accelerant

While acting as a facilitator of the social processes of late-modernity, the internet does more than simply channel or afford such processes. It deepens and intensifies the processes, makes them more evident and accelerates their impact. This occurs in at least three ways.

First, the internet is not the only channel by which such processes are experienced and disseminated. Individualisation is felt or even imposed in many aspects of social life, such as our interactions with institutions and expert systems – social policy and financial systems, family law, trade union recognition and medical insurance systems, for example – all of which impact upon daily life. The internet is a further layer or means by which our experience of individualisation is deepened.

Second, using the internet requires the making of conscious decisions concerning our engagement with others. Our interactions with others and presentation of self are intensified. Over time this subtly forces us to consider the self/group relationship that we live in. The micro-social machinery with which we define ourselves is highlighted and brought to our attention. The point here is not that we choose to engage with such activity, but that we are aware that such activity is now occurring. Even in resisting the pressure to define ourselves online, as some those of a technophobic/neo-Luddite persuasion do, we engage with and recognise the process. To resist we must implicitly be aware of the process we are seeking to resist. The point here is that in resisting we are deepening the process.

Third, our engagement with commercial, state and other institutions through the internet results in greater institutional and macro-level individualisation. In addressing us, such organisational entities make increasing use of techniques to demarcate us from other people, or at least give us the impression that we are being addressed as individuals. We are conceptualised increasingly in smaller and smaller units of measurement. We are targeted, conceived of or understood not as one member of a group but as a member of many different groups. In short, we are instantiated in database systems in an increasingly individualised manner that necessitates our complicity if we wish to engage with the organisation (or in certain instances to resist it).

The internet intensifies the processes of and our experience of late-modernity. It serves to integrate us deeper into late-modern patterns of social life and quicken the impact of such processes. However, the potency of the internet to act in this manner should not be attributed singularly to the internet but to the conditions or environment in which it is deployed.

The contingent nature of the internet

While above I argue that the internet has the power to impact upon social life through its ability to facilitate and accelerate the processes of late-modernity, we also need some way of thinking about how society impacts upon how the internet 'socially' functions. The relationship between media technologies and society is bilateral and operates in two directions: media intersect with society, but the ability of the media to function is tied to specific social formations. As I said above, the internet is an 'active' element of the social world. It makes possible and intersects with behaviour, subtly intensifying the experience of late-modernity. However, the potency of the internet to act in this manner, to change lives, is dependent upon other factors. As the internet is part of society, not external to it, it must be considered in concert with other social activities, processes and elements of social life. That is, we should not consider the media in isolation from other aspects of the social world.

To understand the internet methodologically we must appreciate that it is *contingent* in operation; its propensity to function is dependent upon the correct conditions being in place. The internet can intensify the experience of late-modernity in the manner noted above. More specifically, it can offer new opportunities to voice critical issues, function in a democratically beneficial manner and facilitate new forms of identification, both personal and political. It has in numerous instances operated precisely as described, as a medium that enables dynamic forms of political communication and social change.

However, such properties, forms of action and impact should not be regarded as solely dependent upon the presence of the internet, nor are these forms of action an essence of the internet. Instead, the potency of the internet is contingent; it is dependent upon other factors, a late- or high-modern social form with an increasingly individualised society and a propensity for individual action. This contingency can be thought to function in five 'dimensions'.

First, the extensive *technical systems* that make connection to the internet possible must be in place. This is the most obvious instance of contingency: without technological systems the internet cannot be accessed. By technological systems I refer to the hardware and software being in place so as to make access to the internet possible for a significant number of people. This has many different aspects, ranging from fundamental issues such as the provision of computers of sufficient quality, operating systems and software, electric power and the quality of network connections to more human issues such as technical support for users and server administration. Obvious as they are, such 'hard' issues often receive the most attention and in many instances are the only contingency considered in examining internet penetration.

Second, there must be *institutional systems* to regulate legal and financial issues. There must be laws and financial entities to regulate both user participation and the commercial activity of internet communication. These include the legal instruments that permit the full engagement of users in internet activity, the recognition of copyright and other systems of data protection. They also incorporate a viable economic model for offering internet connectivity. This refers to the commercial (or otherwise) means of offering service to users. Internet provision is a costly service to provide and in virtually all instances provision is made on a commercial basis, with profit being the prime objective of service providers. Accordingly, legal and institutional systems must be in place so as to make profit possible for service providers. While technological systems are essential, they cannot be offered long term without legal and financial institutional systems to support them.

Third, there are *educational issues* – there must be a degree of educational competence in the users. There are at least two forms of knowledge required here. First are the skills needed to access and use internet technologies. These are the basic skills necessary to navigate and find information, and increasingly to be able to produce it. Second, and more politically controversial, is the idea that users need the cultural skills and understanding necessary to interpret much of the material on the internet. This refers to being able to understand how online communication operates, the codes and conventions of interpersonal communication and the ability to understand and critically interpret information. These two sets of skills are often subsumed under the title of media or information literacy, and there is a growing movement advocating the development of such skills and the educational programmes necessary to achieve them. Implicit in such ideas is the recognition that internet presence, the supply of services, is not enough. However, there is still a strong tendency to ignore such 'soft' issues in favour of more concrete issues such as technology availability.

Fourth, related to the point above are *cultural issues;* there must be sufficient culturally specific content available to make use of the internet valuable. If content on the internet is culturally inapplicable then its use serves no purpose in terms of entertainment or otherwise. Such a mismatch between content and user may occur because of linguistic issues or clashes of broader cultural values. While the activity of localisation of web content is a growth industry, what is additionally required is content that is produced specifically for local consumption. This area is a field of key contention in the dissemination of internet provision to developing nations, with strong concerns being raised about the dominance of Western cultural patterns in internet content.

Fifth, there are *sociological issues* – this refers to whether the internet offers something of 'value' to the user. This does not only relate to issues of specific content being of interest, but to more general debates occurring in society. I refer here to the points made above: that the internet seems to facilitate and accelerate the concerns of late-modernity. The internet facilitates those specific debates and issues of late-modernity. If it does not offer issues of interest then it will not function in the manner thought. As explored in Chapter 6, this applied to other media forms in earlier modernity, when newspapers allowed new forms of identity to occur. Media, and in the instance examined here the internet, must offer the type of debate that society is interested in for them to have any impact upon that society.

Unsurprisingly, these five 'dimensions' of contingency are all present in late-modern society. They constitute the conditions that allow the internet to function as it does. Regarding the relationship of the internet to society in this manner does not mean that this is the *only* possible environment in which the internet could function in this manner, but these are the conditions in which it has functioned so far.

Methodological standpoint

With these two arguments we can begin to develop a final methodological stance or standpoint for understanding the relationship of the internet to society. The arguments outlined here result in a particular understanding of the relationship of the internet to late-modern society.

- The internet is an active element of social life and intersects with social systems. The nature of this intersection is that the internet facilitates and accelerates particular processes. However, the propensity to function in this manner is tied to certain conditions; the internet is contingent in its ability to bring about social change.

- Adopting such a position involves 'stepping back' from the direct acceptance of our view of technology or our interpretation of what technology can do. This explicitly challenges the idea that either technology or society should be considered as a priori in conception. Accordingly, analysis will need to accommodate both technology and society equally in the interpretative model. To understand how a media technology operates, we need to examine the relationship between the media technology and the society in which it is studied. It is argued here that the relationship is one of interdependence.

A result of this is that we should move beyond the distinction between 'culturalist' and 'formalist' concerns in attempting to understand the internet. Attention should instead be focused upon the following.

- First, the intersections of technology and of society – the instances of internet use and the social processes and intents to which such use is put. It is at such intersections that the propensity for the internet to enable, facilitate and accelerate social processes is most evident. These intersections occur at a variety of levels: from how we use the internet for various purposes and how the availability of technology modifies social activity to how access to the internet is made possible.

- Second, the conditions required to enable such intersections – the various dimensions of contingency that allow internet activity and the consequent intersections to occur.

Such a 'twin-headed' approach requires us to move our focus beyond technology. It challenges the deterministic vision of technology in which it exists firmly 'outside' society and impacts upon it. In its place we must adopt an integrated approach considering both sociological and technological concerns in equal measure. The model offered here is an attempt to integrate an account of society with a more critically aware understanding of the internet. This is an alternative to models in which human and technological concerns are separated. Such models have long been deployed in justifying communications technology that seemingly benefits a minority of users and corporate interests to the long-term detriment of those who could perhaps benefit most from a more integrated technological and social development programme. As mentioned in Chapter 1, part of the rationale for this book was to contribute to a more progressive account and understanding of the internet. This approach will hopefully be a contribution to the manner in which the internet is understood and utilised, producing a more socially useful account.

In the concluding chapter I examine a number of issues that may be questioned with (or raise questions with) the approach I have suggested here.

Notes

1   It is worth noting that the use of the term 'social system' by Giddens is quite different from that used by more positivist sociologists, who see social systems in terms of the connections between people in networks. It is also different from the same term in economics, where collections of economic relations between agents are similarly termed.

Conclusion

8

Conclusion

Introduction

The intention of this book has been to advance a particular position as to where we should direct our focus when trying to understand the internet and its ability to bring about social change. The idea proposed here is that we look beyond the technology and towards a more integrated account, where the impact of the internet is examined in concert with a 'softer', more sociological analysis. The internet should be examined not on its own, but in the light of and in concert with greater social processes, values, ideals and concerns. Built into this aim was the desire to contribute to a broad school of thought that promotes the idea of democratic policy-making with respect to media technology – the advocacy of a 'bottom-up' approach to its deployment. Perhaps the main contribution that this volume makes to this approach is to stress the importance of 'local' conditions and needs. Appreciating local cultural values, beliefs and conditions is vital in understanding how the technology will impact upon people's lives. Such considerations should be considered paramount in various fields of study that examine how the internet is used and how it impacts upon users. In this final section I wish to address an area for future study and the possible application of the approach suggested here for two fields of study.

Summation of argument

As I mentioned in Chapter 1, the manner in which media technology has been thought to impact upon people's lives has been primarily understood in a simple, linear manner. In the preceding chapters I have sought to show how such an argument is wrong. The impact a 'media technology' has upon a society is complex and multifaceted, and cannot be explained using a simple 'cause-and-effect' model. Instead, I proposed that we recognise that the internet has been imagined through a complex interpretative template. This interpretative template structures our understanding of the internet; we view it using a set of pre-existing interpretative tools. These tools emerge from and reflect the politics and culture of particular societies or countries. I then proposed that the link between a media technology such as the internet and society needs to be considered as 'deeper' and more radical. I argued that we should recognise that the relationship between the internet and society is 'two way' – the internet brings about social change though modifying various social systems so that they exemplify the twin processes of individualisation and detraditionalisation. However, we must also recognise that social conditions, both material and ideational, have considerable bearing upon the impact the internet may have. The internet operates as we think it can because of certain social conditions. These conditions range from hard issues, such as technical and economic infrastructure, to cultural and educational concerns over skills and culturally specific content and the more sociological issues such as the nature of identity formation.

This model problematises the characterisation of the internet as discrete from other aspects of social practice, and requires and allows us to start to think of the internet as a complex element of the social world and not only as one of the technical world.

Key consequence

This book is intended to be a contribution to the sociological study of the internet. It offers an approach that challenges the tradition of regarding technology, and media technology in particular, as external to human life and consequently outside sociological analysis. It is explicitly disputed that media technology and society are separate and discrete. Rather, they are deeply interwoven, operating in a complex, reflective and integrated relationship. Furthermore, when thinking methodologically about where to direct our focus of attention, I argue that we must examine social and technological issues in concert. Neither should be regarded as the 'prime mover' and accorded more importance in explanatory potential. Attention should be paid to both as a collective whole, not as separate, discrete entities.

The key consequence of these ideas is that we develop a more complex interpretation of the relationship of the internet to society. This alternative account will have at its core an appreciation of the potent nature of the internet while at the same time retaining an appreciation of societal and human factors.

Suggestions for further work: research methods

One key area for elaboration that I have not been able to explore here is the development of specific methods and techniques of field enquiry that can be used to investigate the intersections of the internet with social systems. While such activity is really beyond the scope of this book, there are certain methods that would seem to lend themselves to exploring what has been covered.

In this book I have proposed that we extend our analytic 'vision'; that we look beyond issues simply of technology. I argue that we need to integrate accounts of technology into a broader 'social' account in which technological and social processes are not considered discrete, but inherently linked. While this statement is very broad, it can be developed into more concrete proposals. First, as advocated by Miller and Slater (2002), the internet should be examined in its locality of use. This means looking at how the technology 'fits' into daily usage, how it is used in daily life to further the particular aims and intents of its users. Miller and Slater examined how the internet was used by individuals to articulate their sense of being of a particular nationality and community. However, where Miller and Slater adopted an anthropological approach to the internet, I feel that we need to widen the focus slightly to take into account more sociological issues. This means that the focus should be shifted 'upwards' and 'outwards', extending the scope to a greater range. We should be looking at how the internet not only assists the individual but how it intersects with social processes. While Miller and Slater saw how the internet permits individuals to act in a certain manner, what is more interesting (at least for me) is that *it can do this at all*. The ability of the internet to facilitate the act of identity articulation in the manner in which it is widely understood to do so is the phenomenon in question. I argue that it can do so as it intersects with social systems, and modifies them in line with late-modern patterns of identity formation. The internet brings particular social processes 'closer' to the user and increases the individual's experience of them – it plays a considerable part in making us late-modern. At the same time the internet cannot do this alone, and must be considered in concert with other aspects of late-modern life. Significantly, the internet is contingent upon certain social conditions – those of late-modernity. Its ability to bring about change is predicated upon certain conditions or contingencies.

Thus we must understand the internet simultaneously as both an *active force* in society, changing and deepening our experience of late-modernity, but at the same time it is deeply linked to and *dependent* or *contingent* upon the very social conditions it is intensifying. This interpretation requires us to examine society not as a collection of actors with individual interests (though it is of course partly this), but as a collection of social systems through which individuals live. These systems constitute the focal point for the approach advocated here. It is in social systems that the internet impacts upon us – the mid-level practices that allow us to live and bring more macro-level forces to bear upon us. Getting to these systems presents an interesting challenge, as social systems by their very nature are transitory and elusive. They exist not in any material form but in the lived actions of individuals, and consequently they can only be encountered in people's actions. We cannot 'get' to a social system; we can only, at best, see it in operation upon those involved.

The development of actual methods for examining the intersection of the internet with social processes is a distinct area of enquiry that requires further work and is beyond the scope of this book. However, one method that may prove fruitful and has enjoyed considerable success in other fields of enquiry into social systems and processes is the use of a version of 'action research'. Broadly understood, action research is a social research method that seeks reflexively to mix research endeavours with problem-solving or useful (to the participants) activity (Reason and Bradbury, 2001). It integrates the researcher at the core of the observed activity, allowing unparalleled insight into the processes. Action research in which the researcher participates in activity with the observed may offer the best means for researching the direct intersection of the internet with social systems.

Fields of application: information and communication technologies for development

The concern of practitioners in the field of information and communication technologies for development (ICTD) is that of deploying and making available new media, communicational and computing technology with the prime intention of social development. What distinguishes (contemporary) ICTD from other development programmes that use ICT is the advocacy of local interests. Historically, programmes of media technology deployment have been developed and actualised with scant regard for local conditions. They have been developed from 'outside' and have tended to be imposed from 'above', ignoring local needs, conditions and cultural specifics. While an appreciation of local conditions would seem an obvious stepping-stone in media technology deployment for development, it has strangely proven to be a largely ignored area. ICTD has sought to counter this with strong advocacy of the interests of the recipients of technology and factoring 'local' conditions into the deployment of technology.

ICTD's multidisciplinary approach to development – the field brings together technologists, sociologists and development studies experts – allows a perspective that recognises the active interests of users. Furthermore, there has been a gradual shift in ICTD from a position in which ICT deployment was done for the benefit of certain groups to one in which partnership with local groups was considered of key importance to an emerging perspective in which ICT development occurs within and by the communities in question (Heeks, 2008: 29).

The approach advocated here strongly supports such intentions. Technological development and change should be ideally accompanied by and integrated into locally initiated programmes of social, educational and economic development. The potential for or likelihood of beneficial social change occurring following the deployment of new media technology can best be ensured by integrating it into local cultural patterns. Thus programmes of media technological development should be conducted in concert with, not as an alternative to, programmes of educational, cultural, economic and social development. This multifaceted approach is preferable for a number of reasons.

First, factoring in local cultural interests may help to 'normalise' the technology for the intended users. Building new media into already recognisable projects may serve to demystify it and make it more acceptable.

Second, by building the internet into local projects some of the negative or unintended consequences of its content may be mitigated. The current US and Western European dominance of content on the internet and the values associated with such content may be problematic to some. Integrating the internet into existing cultural practices may serve to reduce the negative aspects of this content.

Third, similarly, developing content-production educational programmes that link to cultural practices may assist in the promotion and preservation of existing cultural norms.

Perhaps the strongest reason, however, is that an integrated approach may facilitate development better than a solely technological approach. When media technology is built into local projects, it simply works better. People can see its use and how it can directly benefit their lives.

Fields of application: media studies

In Chapter 4 I mentioned that there is currently a heated debate taking place in media studies concerning whether the subject needs to update itself to accommodate the supposed characteristics of new media. A number of academics have argued that media studies is a set of interpretative tools developed to deal with content mediated through a specific set of technologies. New media, with their different characteristics, necessitate a revision in the tools and frameworks used (Gauntlett, 2000, 2007; Merrin, 2008). In short, media studies should be upgraded to Media Studies 2.0, a new interpretative framework that recognises the qualities within new media and the consequences they may have upon the practice and teaching of media studies.

Criticism of this position has included the argument that technological innovation should not drive social criticism; that much of media studies' power as a critical idiom revolves around its ability to engage with a social agenda, not its dependency upon a technological format. Furthermore, it should not abandon its underlying humanistic and critical tradition, a fear raised by Buckingham (2008). As an academic discipline media studies has a long tradition of challenging orthodox readings of texts and offering radical criticism of key issues, such as gender bias in the media or ownership and control of the media. Such criticism stems from a commitment to radical and progressive politics on the part of practitioners and academics. The fear is that such a moral and critical agenda would be lost or diminished if the interpretative framework is changed.

The ideas presented in this book can be understood to support at least partially the call for a Media Studies 2.0. The challenge posed by new media, with their emphasis upon a more participatory experience for the user, will raise problems for more traditional analytic frameworks. Accordingly, Media Studies 2.0 is concerned with examining the myriad ways in which technologies such as the internet present new challenges that existing frameworks could not address. What I have sought to do here is to suggest one way in which we can go about thinking how the internet intersects with people's lives.

Despite the criticism that Media Studies 2.0 involves abandoning a critical perspective, I argue that it is progressive by its very nature as it places emphasis upon the user. By focusing upon the users' actions, what they are doing with the internet and how it is worked into their lives, attention is drawn to nuances and issues through which power systems affect the individual. While this may not be the macro-level criticism that typifies traditional media studies, it still an inherently political and progressive means of interpretation.

Concluding remarks

The ideas proposed here have sought to challenge the existing interpretation of the internet. I intended to examine or 'tease out' some of the nuances of how the internet has been understood to affect social life. I further wanted to offer a way forward that both acknowledges the common discourses that surround the internet but also explores how use of the internet needs to be understood as being deeply interlinked with social life. The resultant model is a first attempt at this project, and, as with all first attempts, there are many areas in which it may require improvement or revision. This said, I hope I have added a little to the literature in this field.

References

Abbate, J. (2000) *Inventing the Internet*. Cambridge, MA: MIT Press.

Adorno, T. (1991) *The Culture Industry: Selected Essays on Mass Culture*. London: Routledge.

Allen, R.E. (1990) *The Concise Oxford Dictionary*. Oxford: Clarendon Press.

Allender-Hagedorn, S. and Ruggiero, C.W. (2005) 'Connecting popular culture and science: the case of biotechnology', in *Proceedings of the Professional Communication Conference*. Limerick: IEEE, pp. 161–75.

Althusser, L. (1971) 'Ideology and ideological state apparatuses', in L. Althusser (ed.) *Lenin and Philosophy*. New York: Monthly Review Press.

Anderson, B. (1991) *Imagined Communities*. New York: Verso.

Anonymous (2007) 'The third stage and the characteristics of new media', *O'Reilly Commons*; available at: *http://commons.oreilly.com/wiki/index.php/The_third_st age_and_the_characteristics_of_new_media* (accessed: 3 December 2007).

Appadurai, A. (1990) 'Disjuncture and difference in the global cultural economy', *Public Culture* 2(2): 1–24.

Arterton, C. (1987) *Teledemocracy: Can Technology Protect Democracy?* Newbury Park, CA: Sage.

Bagdikian, B. (2004) *The New Media Monopoly*. Boston, MA: Beacon Press.

Barber, B. (1984) *Strong Democracy: Participatory Politics for a New Age*. Berkeley, CA: University of California Press.

Barlow, A. (2007) *The Rise of the Blogosphere*. London: Praeger.

Barlow, J.P. (1996) 'A declaration of the independence of cyberspace'; available at: *http://homes.eff.org/~barlow/Declaration-Final.html* (accessed: 10 April 2008).

Bauman, Z. (1989) *Modernity and the Holocaust*. Cambridge: Polity Press.

Bauman, Z. (2000) *Liquid Modernity*. Cambridge: Polity Press.

Bauman, Z. (2001) *The Individualized Society*. Cambridge: Polity Press.

Bauman, Z. (2002) 'Individually, together', in U. Beck and E. Beck-Gernsheim (eds) *Individualization*. London: Sage.

Bauman, Z. (2004) *Identity: Conversations with Benedetto Vecchi*. Cambridge: Polity Press.

Beck, U. (1992) *Risk Society: Towards a New Modernity*, trans. Mark Ritter. London: Sage.

Beck, U. (1994) 'The reinvention of politics: towards a theory of reflexive modernisation', in U. Beck, A. Giddens and S. Lash (eds) *Reflexive Modernisation: Politics, Tradition and Aesthetics in the Modern Social Order*. Cambridge: Polity Press.

Beck, U. (2007) 'Beyond class and nation: reframing social inequalities in a globalizing world', *British Journal of Sociology*, 58(4): 679–705.

Beck, U. and Beck-Gernsheim, E. (1995) *The Normal Chaos of Love*. Cambridge: Polity Press.

Beck, U. and Beck-Gernsheim, E. (2002) *Individualization*. London: Sage.

Beck, U. and Beck-Gernsheim, E. (2004) 'Families in a runaway world', in J. Scott, J. Treas and M. Richards (eds) *The Blackwell Companion to the Sociology of Families*. Oxford: Blackwell.

Beck, U. and Lau, C. (2005) 'Second modernity as a research agenda: theoretical and empirical explorations in the "meta-change" of modern society', *British Journal of Sociology*, 56(4): 525–57.

Beck, U., Bonß, W. and Lau, C. (2003) 'The theory of reflexive modernization: problematic, hypotheses and research programme', *Theory, Culture and Society*, 20(2): 1–33.

Benhabib, S. (ed.) (1996) *Democracy and Difference: Contesting the Boundaries of the Political*. Princeton, NJ: Princeton University Press.

Berman, M. (1982) *All that Is Solid Melts into Air: The Experience of Modernity*. New York: Penguin.

Bimber, B. (1996) 'Three faces of technological determinism', in M.R. Smith and L. Marx (eds) *Does Technology Drive History? The Dilemma of Technological Determinism*. Cambridge, MA: MIT Press.

Blood, R. (2002) 'Introduction', in J. Rodzvilla (ed.) *We've Got Blog: How Weblogs Are Changing Our Culture*. Cambridge, MA: Perseus Publishing.

Boeder, P. (2005) 'Habermas' heritage: the future of the public sphere in the network society', *First Monday*, 10(9); available at: *http://firstmonday.org/issues/issue10_9/boeder/index.html* (accessed: 26 March 2008).

Bolter, D. (2002) 'Formal analysis and cultural critique in digital media theory', *Convergence*, 8(4): 77–88.

Bordewijk, J.L. and van Kaam, B. (1986) 'Towards a new classification of tele-information services', *Intermedia*, 14(1): 16–21.

Brecht, B. (1979) 'Radio as a means of communication: a talk of the function of radio', in A. Mattelart and S. Siegelaub (eds) *Communication and Class Struggle: 2. Liberation*. New York: International General.

Bryan, C., Tsagarousianou, R. and Tambini, D. (1998) 'Electronic democracy and the civic networking movement in context', in R. Tsagarousianou, D. Tambini and C. Bryan (eds) *Cyberdemocracy: Technology, Cities and Civic Networks*. London: Routledge.

Buckingham, D. (2008) 'Do we really need Media Education 2.0? Prospects for media literacy education in the age of participatory media', keynote address at conference on Digital Content Creation: Creativity, Competence, Critique, University of Southern Denmark, Odense, 18–20 September.

Burawoy, M. (2000) 'Grounding globalization', in M. Burawoy (ed.) *Global Ethnography: Forces, Connections, and Imaginations in a Postmodern World*. Berkeley, CA: University of California Press.

Butler, J. (1990) *Gender Trouble: Feminism and the Subversion of Identity*. London: Routledge.

Butler, J. (1993) *Bodies that Matter: On the Discursive Limits of 'Sex'*. London: Routledge.

Butler, J. (1997a) *Excitable Speech: A Politics of the Performative*. London: Routledge.

Butler, J. (1997b) *The Psychic Life of Power: Theories of Subjection*. Stanford, CT: Stanford University Press.

Calhoun, C. (ed.) (1992) *Habermas and the Public Sphere*. Cambridge, MA: MIT Press.

Callon, M. (1991) 'Techno-economic networks and irreversibility', in J. Law (ed.) *A Sociology of Monsters:*

Essays on Power, Technology and Domination. London and New York: Routledge, pp. 132–61.

Cammaerts, B. (2007) 'Citizenship, the public sphere, and media', in B. Cammaerts and N. Carpentier (eds) *Reclaiming the Media: Communication Rights and Democratic Media Roles.* Bristol: Intellect.

Carey, J.W. (1989) *Communication as Culture: Essays on Media and Society.* London: Unwin Hyman.

Carey, J.W. (2005) 'Historical pragmatism and the internet', *New Media and Society,* 7(4): 443–55.

Carlyle, T. (1859) *On Heroes: Hero Worship and the Heroic in History: Six Lectures Reported with Emendations and Additions.* New York: Wiley and Halstead.

Castells, M. (1996) *The Rise of the Network Society.* Oxford: Blackwell.

Castronova, E. (2006) *Synthetic Worlds: The Business and Culture of Online Games.* Chicago, IL: University of Chicago Press.

Castronova, E. (2007) *Exodus to the Virtual World: How Online Fun Is Changing Reality.* New York: Palgrave Macmillan.

Cavalier, R.J. (2005) *The Impact of the Internet on Our Moral Lives.* New York: SUNY Press.

Chapman, N. and Chapman, J. (2000) *Digital Multimedia.* Brisbane: Wiley.

Chase, S. (1929) *Men and Machines.* New York: Macmillan.

Cole, J., Suman, M., Schramm, P., Lunn, R., Coget, J.F., Firth, D., Fortier, D., Hanson, K., Qin, J., Singh, R., Yamauchi, Y. and Aquino, J.S. (2001) 'The UCLA Internet Report 2001: surveying the digital future, year two'; available at: *http://ccp.ucla.edu/pdf/UCLA-Internet-Report-2001.pdf* (accessed: 17 September 2003).

Colvile, R. (2008) *Politics, Policy and the Internet.* London: Centre for Policy Studies.

Corner, J. (2000) 'Influence: the contested core of media research', in J. Curran and M. Gurevitch (eds) *Mass Media and Society*, 3rd edn. London: Arnold.

Crawford, C. (1990) 'Lessons from computer games design', in B. Laurel (ed.) *The Art of Human-Computer Interface Design*. Reading, MA: Addison-Wesley.

Crawford, C. (2002) *Understanding Interactivity*. San Francisco, CA: No Starch Press.

Crystal, D. (2006) *Language and the Internet*. Cambridge: Cambridge University Press.

Curran, J. (1991) 'Rethinking the media as a public sphere', in P. Dahlgreen and C. Sparks (eds) *Communication and Citizenship: Journalism and the Public Sphere in the New Media Age*. London: Routledge.

Curran, J. (2000) 'Rethinking media and democracy', in J. Curran and M. Gurevitch (eds) *Mass Media and Society*, 3rd edn. London: Arnold.

Dahlberg, L. (1998) 'Cyberspace and the public sphere', *Convergence*, 4(1): 70–84.

Dahlberg, L. (2001a) 'Computer-mediated communication and the public sphere: a critical analysis', *Journal of Computer-Mediated Communication*, 7(1); available at: *http://jcmc.indiana.edu/vol7/issue1/dahlberg.html* (accessed: 26 March 2008).

Dahlberg, L. (2001b) 'Democracy via cyberspace', *New Media and Society*, 3(2): 157–77.

Dahlberg, L. (2001c) 'The internet and democratic discourse: exploring the prospects of online deliberative forums extending the public sphere', *Information, Communication and Society*, 4(4): 615–33.

Dahlberg, L. (2005) 'The corporate colonization of online attention and the marginalization of critical communication', *Journal of Communication Inquiry*, 29(2): 1–21.

Dahlberg, L. (2007) 'Rethinking the fragmentation of the cyberpublic: from consensus to contestation', *New Media and Society*, 9(5): 827–47.

Dahlberg, L. and Siapera, E. (eds) (2006) *Radical Democracy and the Internet: Interrogating Theory and Practice*. Basingstoke: Palgrave Macmillan.

Dawkins, R. (1989) *The Selfish Gene*. Oxford: Oxford University Press.

DeFleur, M. and Ball-Rokeach, S. (1989) *Theories of Mass Communication*. New York: Longman.

Dertouzos, M. (1991) 'Communication, computers and networks', *Scientific America*, 256(3): 74–80.

Dovey, J. (2002) 'Intertextual tie ups: when narratology met Ludology', paper presented at Playing with the Future: Development and Directions in Computer Gaming conference, University of Manchester, 5–7 April.

Downes, E. and McMillan, S. (2000) 'Defining interactivity: a qualitative identification of key dimensions', *New Media and Society*, 2(2): 157–79.

Downey, J. and Fenton, N. (2003) 'New media, counter publicity and the public sphere', *New Media and Society*, 5(2): 185–202.

Downing, J. (1996) *Internationalising Media Theory*. London: Sage.

Downing, J. (2001) *Radical Media: Rebellious Communication and Social Movements*. London: Sage.

Dreyfus, H. (2001) *On the Internet*. London: Routledge.

Dusek, V. (2006) *Philosophy of Technology: An Introduction*. Malden, MA: Blackwell.

Elster, J. (1985) *Making Sense of Marx*. Cambridge: Cambridge University Press.

Enzensberger, H.M. (1970) 'Constituents of a theory of the media', *New Left Review*, 64: 13–36.

Escobar, A. (1994) 'Welcome to Cyberia: notes on the anthropology of cyberculture (and comments and reply)', *Current Anthropology*, 35(3): 211–31.

Esteva, G. and Prakash, M.S. (1998) *Grassroots Postmodernism: Remaking the Soil of Cultures*. London: Zed Books.

Feenberg, A. (1999a) *Questioning Technology*. London: Routledge.

Feenberg, A. (1999b) 'Modernity theory and technology studies: reflections on bridging the gap', paper presented at Technology and Modernity Conference, University of Twente, November; available at: *www-rohan.sdsu.edu/faculty/feenberg/twente.html* (accessed: 11 December 2007).

Feenberg, A. (2003) 'What is philosophy of technology?'; available at: *www-rohan.sdsu.edu/faculty/feenberg/komaba.htm* (accessed: 11 December 2007).

Feenberg, A. (2004) 'Modernity theory and technology studies', in T. Misa, P. Bray and A. Feenberg (eds) *Modernity and Technology*. Cambridge, MA: MIT Press.

Felski, R. (1997) 'The Doxa of difference', *Signs: Journal of Women in Culture and Society*, 23(1): 1–23.

Ferguson, M. (1991) 'Marshall McLuhan revisited: 1960s zeitgeist victim or pioneer postmodernist?', *Media, Culture and Society*, 13: 71–90.

Fraser, N. (1992a) 'Rethinking the public sphere: a contribution to the critique of actually existing democracy', in C. Calhoun (ed.) *Habermas and the Public Sphere*. Cambridge, MA: MIT Press.

Fraser, N. (1992b) 'Sex, lies, and the public sphere: some reflections on the confirmation of Clarence', *Critical Theory*, 18(3): 595–612.

Freeman, C. (1992) 'The case for technological determinism', in R. Finnigan, G. Salamam and K.

Thompson (eds) *Information Technology: Social Issues – A Reader*. London: Hodder & Stoughton.

Froomkin, M. (2003) 'Habermas@Discourse.Net: toward a critical theory of cyberspace', *Harvard Law Review*, 16(3): 749–873.

Gackenbach, J. (ed.) (2006) *Psychology and the Internet: Intrapersonal, Interpersonal, and Transpersonal Implications*, 2nd edn. Orlando, FL: Academic Press.

Gates, B. (2000) *Business at the Speed of Thought: Using a Digital Nervous System*. London: Penguin.

Gauntlett, D. (2000) 'Web studies: a user's guide', in D. Gauntlett (ed.) *Web.Studies: Rewiring Media Studies for the Digital Age*. London: Edward Arnold.

Gauntlett, D. (2007) 'Media Studies 2.0'; available at: *www.theory.org.uk/mediastudies2.htm* (accessed: 1 April 2008).

Gehlen, A. (1983) 'A philosophical-anthropological perspective on technology', *Research in Philosophy and Technology*, 6: 205–16.

Gellner, E. (1964) *Thought and Change*. London: Weidenfeld & Nicolson.

Gellner, E. (1983) *Nations and Nationalism*. Oxford: Blackwell.

Gellner, E. (1997) *Nationalism*. London: Weidenfeld & Nicolson.

Giddens, A. (1976) *New Rules of Sociological Method*. New York: Basic Books.

Giddens, A. (1979) *Central Problems in Social Theory: Action, Structure and Contradiction in Social Analysis*. London: Macmillan.

Giddens, A. (1984) *The Constitution of Society. Outline of the Theory of Structuration*. Cambridge: Polity Press.

Giddens, A. (1990) *The Consequences of Modernity*. Cambridge: Polity Press.

Giddens, A. (1991) *Modernity and Self-Identity*. Cambridge: Polity Press.

Giddens, A. (1994) 'Risk, trust, reflexivity', in U. Beck, A. Giddens and S. Lash (eds) *Reflexive Modernisation: Politics, Tradition and Aesthetics in the Modern Social Order*. Cambridge: Polity Press.

Gilder, G. (1994) *Life after Television: The Coming Transformation of Media and American Life*. New York: W.W. Norton.

Gilmor, D. (2006) *We the Media: Grassroots Journalism by the People for the People*. Sebastopol, CA: O'Reilly Media.

Gimmler, A. (2001) 'Deliberative democracy, the public sphere and the internet', *Philosophy and Social Criticism*, 27(4): 21–39.

Golding, P. and Murdock, G. (2000) 'Culture, communications and political economy', in J. Curran and M. Gurevitch (eds) *Mass Media and Society*, 3rd edn. London: Arnold.

Gomez, J. (2004) 'Dumbing down democracy: trends in internet regulation, surveillance and control in Asia', *Pacific Journalism Review*, 10(2): 130–50.

Gorard, S. and Selwyn, N. (2001) *101 Key Ideas in Information Technology*. London: Hodder & Stoughton.

Gore, A. (1994) 'The global information infrastructure: forging a new Athenian age of democracy', *Intermedia*, 22(2): 4–7.

Graham, G. (1999) *The Internet:// A Philosophical Inquiry*. London: Routledge.

Gramsci, A. (1971) *Selection from Prison Notebooks*, trans. Q. Hoare and G. Nowell-Smith. London: Lawrence and Wishart.

Habermas, J. (1985) *Theory of Communicative Action Vol. 1: Reason and the Rationalisation of Society*, trans. T. McCarthy. Boston, MA: Beacon Press.

Habermas, J. (1989) *The Structural Transformation of the Public Sphere: An Inquiry into a Category of Bourgeois Society*, trans. T. Burger with F. Lawrence. Cambridge: Polity Press.

Habermas, J. (1992) 'Further reflections on the public sphere', in C. Calhoun (ed.) *Habermas and the Public Sphere*. Cambridge, MA: MIT Press.

Habermas, J. (1996) *Between Facts and Norms: Contributions to a Discourse Theory of Law and Democracy*, trans. W. Rehg. Cambridge: Polity Press.

Habermas, J. (2004) 'Public space and political public space – the biographical roots of two motifs in my thoughts', Kyoto Prize speech; available at: *http://homepage.mac.com/gedavis/JH/Kyoto_lecture_Nov_2004.pdf* (accessed: 27 March 2008).

Habermas, J. (2005) 'Concluding comments on empirical approaches to deliberative politics', *Acta Politica*, 40(3): 384–92.

Habermas, J. (2006) 'Political communication in media society: does democracy still enjoy an epistemic dimension? The impact of normative theory on empirical research', *Communication Theory*, 16(4): 411–26.

Hall, P. and Soskice, D. (2001) *Varieties of Capitalism: The Institutional Foundations of Comparative Advantage*. Oxford: Oxford University Press.

Hall, S. (1981) 'Encoding/decoding in TV discourse', in Centre for Contemporary Cultural Studies (ed.) *Culture, Media, Language*. London: Hutchinson.

Hall, S. and Jefferson, S. (1976) *Resistance through Rituals: Youth Sub-cultures in Post-war Britain*. London: Unwin Hyman.

Hamlink, C. (1995) *World Communication: Disempowerment and Self-Empowerment*. London: Zed Books.

Hård, M. and Jamison, A. (1998) 'Conceptual framework: technology debates as appropriation processes', in M. Hård and A. Jamison (eds) *The Intellectual Appropriation of Technology: Discourses on Modernity 1900–1930*. Cambridge, MA: MIT Press.

Harding, J. (1998) *Sex Acts: Practices of Femininity and Masculinity*. London: Sage.

Hargittai, E. (2000) 'Open portals or closed gates? Channeling content on the World Wide Web', *Poetics*, 27(4): 233–54.

Hargittai, E. (2007) 'The social, political, economic, and cultural dimensions of search engines: an introduction', *Journal of Computer-Mediated Communication*, 12(3); available at: *http://jcmc.indiana.edu/vol12/issue3/hargittai.html* (accessed: 12 February 2008).

Harindranath, R. (2006) *Perspectives on Global Cultures*. Maidenhead: Open University Press.

Harraway, D. (1991) *Simians, Cyborgs and Women: The Reinvention of Nature*. London: Free Association.

Heeks, R. (2008) 'ICT4D 2.0: The next phase of applying ICT for international development', *Computer*, 41(6): 26–33; available at: *http://csdl2.computer.org/comp/mags/co/2008/06/mco2008060026.pdf* (accessed: 19 September 2008).

Heelas, P. (1996) 'Introduction: detraditionalization and its rivals', in P. Heelas, S. Lash and P. Morris (eds) *Detraditionalization: Critical Reflections on Authority and Identity at a Time of Uncertainty*. Oxford: Blackwell.

Heidegger, M. (1977) *The Question Concerning Technology*, trans. W. Lovitt. New York: Harper Torch Books.

Heilbroner, R. (1996) 'Technological determinism revisited', in M.R. Smith and L. Marx (eds) *Does Technology Drive History? The Dilemma of Technological Determinism.* Cambridge, MA: MIT Press.

Held, D. and Thompson, J. (1989) 'Editors' introduction', in D. Held and J. Thompson (eds) *Social Theory of Modern Societies: Anthony Giddens and His Critics.* Cambridge: Cambridge University Press.

Herman, E. and Chomsky, N. (1988) *Manufacturing Consent: The Political Economy of the Mass Media.* New York: Pantheon.

Hewitt, H. (2006) *Blog: Understanding the Information Reformation That's Changing Your World.* Nashville, TN: Nelson Business.

Hill, K.A. and Hughes, J.E. (1998) *Cyberpolitics: Citizen Activism in the Age of the Internet.* Lanham, MD: Rowman & Littlefield.

Hine, C. (2000) *Virtual Ethnography.* London: Sage.

Holland, N.N. (1996) 'The internet regression'; available at: *www-usr.rider.edu/~suler/psycyber/holland.html* (accessed: 12 September 2007).

Jenkins, H. (2006) *Fans, Bloggers, and Gamers: Exploring Participatory Culture.* New York: New York University Press.

Jensen, J.F. (1998) '"Interactivity". Tracking a new concept in media and communication studies', *Nordicom Review*, 19(1): 185–204; available at: *www.nordicom.gu.se/ reviewcontents/ncomreview/ncomreview198/jensen.pdf* (accessed: 23 January 2008).

Joinson, A. (2003) *Understanding the Psychology of Internet Behaviour: Virtual Worlds, Real Lives.* New York: Palgrave Macmillan.

Jones, P. (1998) 'The technology is not the cultural form? Raymond Williams's sociological critique of Marshall

McLuhan', *Canadian Journal of Communication*, 23(4); available at: *www.cjc-online.ca/viewarticle.php ?id=481andlayout=html* (accessed: 26 November 2007).

Jones, S. (1998) *Doing Internet Research: Critical Issues and Methods for Examining the Net*. Newbury Park, CA: Sage.

Kaya, I. (2004) 'Modernity, openness, interpretation: a perspective on multiple modernities', *Social Science Information*, 43(1): 35–57.

Keane, J. (1991) *The Media and Democracy*. Cambridge: Polity Press.

Keane, J. (2000) 'Structural transformations of the public sphere', in Jan van Djik and K. Hacker (eds) *Digital Democracy; Issues of Theory and Practice*. London: Sage.

Kedourie, E. (1993) *Nationalism*. Oxford: Blackwell.

Kellas, J. (1991) *The Politics of Nationalism and Ethnicity*. Basingstoke: Macmillan.

Kellner, D. (1998) 'Intellectuals, the new public spheres, and techno-politics', in C. Toulouse and T. Luke (eds) *The Politics of Cyberspace*. New York: Routledge.

Kendall, G. and Wickham, G. (1999) *Using Foucault's Methods*. London: Sage.

Keren, M. (2006) *Blogosphere: The New Political Arena*. Lanham, MD: Lexington Books.

Kline, D. (2005) 'Toward a more participatory democracy', in D. Kline and D. Burstein (eds) *Blog!: How the Newest Media Revolution Is Changing Politics, Business, and Culture*. New York: CDS.

Kline, S. (1985) 'What is technology?', *Bulletin of Science, Technology and Society*, 1: 215–18.

Kollock, P. and Smith, M. (2001) 'Communities in cyberspace', in P. Kollock and M. Smith (eds) *Communities in Cyberspace*. London: Routledge.

Kolo, C. and Baur, T. (2004) 'Living a virtual life: social dynamics of online gaming', *Games Studies*, 4(1); available at: *www.gamestudies.org/0401/kolo/* (accessed: 5 February 2008).

Kuhn, T. (1962) *The Structure of Scientific Revolutions*. Chicago, IL: Chicago University Press.

Laclau, E. (1990) *New Reflections on the Revolution in Our Time*. New York: Verso.

Laclau, E. (2005) *On Populist Reason*. New York: Verso.

Laclau, E. and Mouffe, C. (1985) *Hegemony and Socialist Strategy*. London: Verso.

Laclau, E. and Zac, L. (1994) 'Minding the gap: the subject of politics', in E. Laclau (ed.) *The Making of Political Identities*. London: Verso.

Lash, S. (1987) 'Modernity or modernism: Weber and contemporary social theory', in S. Wimster and S. Lash (eds) *Max Weber, Rationality and Modernity*. London: Allen & Unwin.

Lash, S. (2001) 'Individualization in a non-linear mode', in U. Beck and E. Beck-Gernsheim (eds) *Individualization*. London: Sage.

Lasica, J.D. (2002a) 'The second coming of personalized news', *USC Annenberg Online Journalism Review: Technology*; available at: *www.ojr.org/ojr/lasica/ 1017779244.php* (accessed: 18 December 2007).

Lasica, J.D. (2002b) 'The promise of the Daily Me', *USC Annenberg Online Journalism Review: Technology*; available at: *www.ojr.org/ojr/technology/1017778824.php* (accessed: 18 December 2007).

Latour, B. (1988) 'Mixing humans and nonhumans together: the sociology of a door closer', *Social Problems*, 35(3): 298–310.

Latour, B. and Woolgar, S. (1986) *Laboratory Life: The Construction of Scientific Facts*. Princeton, NJ: Princeton University Press.

Levy, A. (2006) *Female Chauvinist Pigs*. New York: Free Press.

Lister, M., Kelly, K., Dovey, J., Giddings, S. and Grant, I. (2002) *New Media: A Critical Introduction*. London: Routledge.

Logan, R. (2007) 'The 14 messages of new media'; available at: *www.pbs.org/mediashift/2007/08/extending_mcluhanthe_14_messag.html* (accessed: 8 September 2008).

Lyon, D. (1988) *The Information Society: Issues and Illusions*. Malden, MA: Polity Press.

Manovich, L. (2001) *The Language of New Media*. Cambridge, MA: MIT Press.

Marcuse, H. (1968) *Negations*, trans. J. Shapiro. London: Penguin.

Marcuse, H. (1991) *One Dimensional Man*. London: Routledge.

Marvin, C. (1990) *When Old Technologies Were New*. Oxford: Oxford University Press.

Marx, K. (1955) *The Poverty of Philosophy*. Moscow: Progress Publisher.

Marx, L. and Smith, M. (1996) 'Introduction', in M. Smith and L. Marx (eds) *Does Technology Drive History? The Dilemma of Technological Determinism*. Cambridge, MA: MIT Press.

McCarthy, T. (1994) 'Introduction', in J. Habermas, *The Structural Transformations of the Public Sphere*, 2nd edn, trans. J. McCarthy. Cambridge: Polity Press.

McLuhan, M. (1962) *The Gutenberg Galaxy: The Making of Typographic Man*. Toronto: University of Toronto Press.

McLuhan, M. (1964) *Understanding Media: The Extensions of Man*. New York: McGraw-Hill.

McLuhan, M. and Fiore, Q. (1967) *The Medium is the Massage*. New York: Bantam Books.

McLuhan, M. and Powers, B. (1989) *The Global Village: Transformations in World Life and Media in the 21st Century*. Oxford: Oxford University Press.

McQuail, D. (1986) 'Is media theory adequate to the challenges of new communications technology?', in M. Ferguson (ed.) *New Communications Technologies and the Public Interest: Comparative Perspectives on Policy and Research*. London: Sage.

McQuail, D. (2001) 'Emerging challenges to media theory', in S. Melkote and S. Rao (eds) *Critical Issues in Communication*. London: Sage, pp. 289–306.

Melucci, A. (1980) 'The new social movements: a theoretical approach', *Social Science Information*, 19(2): 199–226.

Merrin, W. (2008) 'Media Studies 2.0 – my thoughts...'; available at: *http://mediastudies2point0.blogspot com/2008/01/for-number-of-years-ive-been-thinking.html.* (accessed: 23 January 2008).

Miller, D. and Slater, D. (2000) *The Internet: An Ethnographic Approach*. London: Berg.

Miller, E. (2003) 'What does technology mean?'; available at: *www.soundscapes.dk/What%20Does%20technology %20mean.doc* (accessed: 25 August 2003).

Modelski, T. (1986) *Studies in Entertainment: Critical Approaches to Mass Culture*. Bloomington, IN: Indiana University Press.

Morris, M. and Ogan, C. (1997) 'The internet as a mass medium', *Journal of Computer-Mediated Communication*, 1(4); available at: *http://jcmc.indiana. edu/vol1/issue4/morris.html* (accessed: 3 December 2007).

Mouffe, C. (1999) 'Deliberative democracy or agonistic pluralism?', *Social Research*, 66(3): 746–58.

Multimedia Development Corporation (2000) 'About MSC – overview'; available at: *www.msc.com.my/mdc/msc/default.asp* (accessed: 13 January 2002).

Murdock, G. and Golding, P. (1977) 'Capitalism, communication and class relations', in J. Curran and M. Gurevitch (eds) *Mass Media and Society*. London: Arnold.

Napoli, P.M. (2003) *Audience Economics, Media Institutions and the Audience Marketplace*. New York: Columbia University Press.

Negroponte, N. (1996) *Being Digital*. London: Hodder & Stoughton.

Newbold, C., Boyd-Barrett, O. and Van den Bulck, H. (2002) *The Media Book*. London: Hodder Arnold.

Nowell-Smith, G. (1991) 'Broadcasting: national cultures/international business', *New Formations*, 13 (Spring): 39–44.

Ó Baoill, A. (2005) 'Weblogs and the public sphere', in L. Gurak, S. Antonijevic, L. Johnson, C. Ratliff and J. Reyman (eds) *Into the Blogosphere: Rhetoric, Community and the Culture of Weblogs*; available at: *http://blog.lib.umn.edu/blogosphere/weblogs_and_the_p ublic_sphere.html* (accessed: 20 February 2008).

Paul, P. (2005) *Pornified: How Pornography Is Transforming Our Lives, Our Relationships, and Our Families*. New York: Times Books.

Pope, A. (1870) 'An essay on man', in H.F. Cary (ed.) *Poetical Works*. London: Routledge.

Poster, M. (1995) *CyberDemocracy: Internet and the Public Sphere*; available at: *www.hnet.uci.edu/mposter/writings/democ.html* (accessed: 4 March 2008).

Poster, M. (1996) *The Second Media Age*. Cambridge: Polity Press.

Poster, M. (1997) 'Postmodern virtualities', in M. Featherstone and R. Burrows (eds) *Cyberspace, Cyberbodies, Cyberpunk: Cultures of Technological Embodiment*. London: Sage.

Postman, N. (1993) *Technopoly: The Surrender of Culture to Technology*. New York: Vintage.

Poulet, G. (1966) 'Romanticism', in A.K. Thorlby (ed.) *The Romantic Movement*. London: Longman.

Press, A. (2000) 'Recent developments in feminist communication theory: difference, public sphere, body and technology', in J. Curran and M. Gurevitch (eds) *Mass Media and Society*, 3rd edn. London: Arnold.

Qvortrup, L. (1984) *The Social Significance of Telematics: An Essay on the Information Society*, trans. Phillip Edmonds. Philadelphia, PA: John Benjamins.

Raymond, E.S. (1997) *The Cathedral and the Bazaar*. Sebastopol, CA: O'Reilly and Associates.

Reason, P. and Bradbury, H. (eds) (2001) *Handbook of Action Research*. Thousand Oaks, CA: Sage.

Rheingold. H. ([1993] 2000) *Virtual Reality*. Cambridge, MA, and London: MIT Press.

Rheingold, H. (2008) 'Habermas blows off question about the internet and the public sphere'; available at: *www.smartmobs.com/2007/11/05/habermas-blows-off-question-about-the-internet-and-the-public-sphere/* (accessed: 12 March 2008).

Rice, R. and Williams, F. (1984) 'New media technology: the study of new media', in R. Rice (ed.) *Communication, Research, and Technology*. Beverly Hills, CA: Sage.

Rogers, E.M. (1986) *Communication Technology: The New Media in Society*. New York: Free Press.

Rogers, E.M. and Chaffee, S. (1983) 'Communication as an academic discipline: a dialogue', *Journal of Communication*, 33(3): 18–30.

Sassi, S. (2001) 'The transformation of the public sphere', in B. Axford and R. Huggins (eds) *New Media and Politics*. London: Sage.

Scannell, P. (1988) 'Radio times: the temporal arrangements of broadcasting in the modern world', in P. Drummond and R. Paterson (eds) *Television and its Audience*. London: BFI.

Schiller, H. (1969) *Mass Communications and American Empire*. New York: A.M. Kelley.

Schiller, H. (1976) *Communication and Cultural Domination*. New York: International Arts and Sciences Press.

Schlesinger, P. (1991) *Media, State and Nation: Political Violence and Collective Identities*. London: Sage.

Schmidt, V. (2006) 'Multiple modernities or varieties of modernity?', *Current Sociology*, 54(1): 77–97.

Sclove, R. (1995) *Democracy and Technology*. New York: Guilford Press.

Scott-Hall, R. (2006) *Blog Ahead, The: How Citizen Generated Media Is Tilting the Communications Balance*. New York: Morgan James.

Selnow, G.W. (1998) *Electronic Whistle Stops: The Impact of the Internet on American Politics*. Westport, CT: Praeger.

Selwyn, N. (2004) 'Reconsidering political and popular understandings of the digital divide', *New Media and Society*, 6(3): 341–62.

Selwyn, N. and Gorard, S. (2002) *The Information Age: Technology, Learning and Exclusion in Wales*. Cardiff: University of Wales Press.

Shneiderman, B. (1997) *Designing the User Interface: Strategies for Effective Human-Computer Interaction.* Boston, MA: Addison-Wesley.

Slevin, J. (2000) *The Internet and Society.* Cambridge: Polity Press.

Smith, A.M. (1994) 'Rastafari as resistance and the ambiguities of essentialism in the "new social movements"', in E. Laclau (ed.) *The Making of Political Identities.* London: Verso.

Spender, D. (1995) *Nattering on the Net.* Melbourne: Spinifex.

Sreberny-Mohammadi, A. and Mohammadi, A. (1994) 'Small media and revolutionary change: a new model', in A. Sreberny-Mohammadi, D. Winseck, J. McKenna and O. Boyd-Barrett (eds) *Media in Global Context: A Reader.* London: Hodder Arnold.

St Amant, K. (2007) *Linguistic and Cultural Online Communication: Issues in the Global Age.* Hershey, PA: IGI Global.

Standage, T. (1999) *The Victorian Internet.* London: Phoenix.

Steemers, J. (ed.) (1998) *Changing Channels: The Prospects for Television in a Digital World.* Luton: University of Luton Press.

Stoll, C. (1995) *Silicon Snake Oil.* New York: Doubleday.

Strinarti, D. (1995) *An Introduction to Theories of Popular Culture.* London: Routledge.

Stromer-Galley, J. (2003) 'Diversity of political conversation on the internet: users' perspectives', *Journal of Computer-Mediated Communication,* 8(3); available at: *http://jcmc.indiana.edu/vol8/issue3/stromergalley.html* (accessed: 28 March 2008).

Sunstein, C. (2001) *Republic.com.* Princeton, NJ: Princeton University Press.

Surry, D. and Farquhar, J. (1997) 'Diffusion theory and instructional technology', *Journal of Instructional Science and Technology*, 2(1); available at: *www.usq.edu.au/electpub/e-jist/docs/old/vol2no1/article2.htm* (accessed: 11 December 2007).

Tambini, D., Leonardi, D. and Marsden, C. (2007) *Codifying Cyberspace: Communications Self-Regulation in the Age of Internet Convergence*. London: Routledge.

Telesin, J. (1973) 'Inside Samizdat', *Encounter*, 40(2): 25–33.

Terzis, G. (2008) *European Media Governance: The National and Regional Dimensions*. Bristol: Intellect.

Thompson, A. and Fevre, R. (2001) 'The national question: sociological reflections on nations and nationalism', *Nations and Nationalism*, 7(3): 297–316.

Thompson, J. (1990) *Ideology and Modern Culture*. Cambridge: Polity Press.

Thompson, J. (1995) *Media and Modernity*. Cambridge: Polity Press.

Toffler, A. (1971) *Future Shock*. New York: Bantam Books.

Torfing, J. (1999) *New Theories of Discourse*. Oxford: Blackwell.

Tsagarousianou, R. (1998) 'Electronic democracy and the public sphere: opportunities and challenges', in R. Tsagarousianou, D. Tambini and C. Bryan (eds) *Cyberdemocracy: Technology, Cities and Civic Networks*. London: Routledge.

Tuchman, G. (1978) 'The symbolic annihilation of women by mass media', in A.K. Daniels and J. Benet (eds) *Hearth and Home: Images of Women in the Mass Media*. New York: Oxford University Press.

Turkle, S. (1995) *Life on the Screen*. London: Weidenfeld & Nicholson.

Turner, B. (1990) 'Periodization and politics in the postmodern', in B. Turner (ed.) *Theories of Modernity and Postmodernity*. London: Sage.

Turner, F. (2006) *From Counterculture to Cyberculture: Stewart Brand, the Whole Earth Network, and the Rise of Digital Utopianism*. Chicago, IL: University of Chicago Press.

Turner, G. (1996) *British Cultural Studies: An Introduction*, 2nd edn. London: Unwin Hyman.

van Dijk, J. (2006) *The Network Society*, 2nd edn. London: Sage.

van Zoonen, L. (1984) *Feminist Media Studies*. London: Sage.

van Zoonen, L. (1991) 'Feminist perspectives on the media', in J. Curran and M. Gurevitch (eds) *Mass Media and Society*, 2nd edn. London: Arnold.

Villarreal Ford, T. and Gil, G. (2000) 'Radical internet use', in J. Downing (ed.) *Radical Media: Rebellious Communication and Social Movements*. London: Sage.

Walch, J. (1999) *In the Net*. London: Zed Books.

Wallace, P. (2001) *The Psychology of the Internet*. Cambridge: Cambridge University Press.

Waugh, P. (ed.) (1992) *Postmodernism: A Reader*. London: Edward Arnold.

Welch, K. (1999) *Electric Rhetoric: Classical Rhetoric, Oralism, and a New Literacy*. Cambridge, MA: MIT Press.

Wellman, B., Quan, A., Witte, J. and Hampton, K. (2002) 'Capitalizing on the internet: network capital, participatory capital, and a sense of community', in B. Wellman and C. Haythornthwaite (eds) *The Internet and Everyday Life*. Oxford: Blackwell.

Welsh Assembly Government (undated) 'Online for a better Wales'; available at: *http://wales.gov.uk/news/*

*archivepress/enterprisepress/einpress2001/749982/?lang=
en* (accessed: 11 December 2007).

Wilhelm, A.G. (1999) 'Virtual sounding boards: how deliberative is online political discussion?', in B.N. Hague and B.D. Loader (eds) *Digital Democracy: Discourse and Decision Making in the Information Age*. London: Routledge.

Wilhelm, A.G. (2000) *Democracy in the Digital Age*. New York: Routledge.

Williams, F., Rice, R. and Rogers, E.M. (1988) *Research Methods and the New Media*. New York: Free Press.

Williams, R. (1974) *Television: Technology and Cultural Form*. London: Fontana.

Williams, R. (1981) *Culture*. London: Fontana.

Winner, L. (1987) *Autonomous Technology: Technics-out-of-Control as a Theme in Political Thought*. Cambridge, MA, and London: MIT Press.

Winner, L. (1996) 'Do artefacts have politics?', in D. MacKenzie and J. Wajcman (eds) *The Social Shaping of Technology*. Buckingham: Open University Press.

Winston, B. (1998) *Media Technology and Society, a History: From the Telegraph to the Internet*. London: Routledge.

Wong, K. and Sayo, P. (2004) *Free Open Source Software: A General Introduction*. Kuala Lumpur: UN Development Programme/Asia-Pacific Development Information Programme (UNDP-APDIP)

Wortham, J. (2007) 'After 10 years of blogs, the future's brighter than ever', *Wired*; available at: *www.wired.com/entertainment/theweb/news/2007/12/blog_anniversary* (accessed: 12 February 2008).

Wright-Mills, C. (1956) *The Power Elite*. New York: Oxford University Press.

Yamamura, K. and Streeck, W. (eds) (2003) *The End of Diversity? Prospects for German and Japanese Capitalism.* Ithaca, NY: Cornell University Press.

Index

Printed and bound by CPI Group (UK) Ltd, Croydon, CR0 4YY

08/10/2024

01042298-0001